The First Passive Solar Catalog

David A. Bainbridge

ISBN O-933490-00-3

Library of Congress No. 78-78403

The Passive Solar Institute
P.O. 722
Davis, CA 95616

a) How to Use This Book

Passive solar design can make your building more economical to build and operate, more comfortable in winter and summer, more secure, and healthier. This book should help you design and build your own passive solar space heating and cooling systems, water heaters, and natural lighting systems.

This book can be read in five ways depending on your interest and hurry.

First, for access to passive solar components go directly to Chapter 11, part 1.

Second, for access to passive solar consultants, go directly to Chapter 11, part 2.

Third, as an introduction to the foundation of passive solar design and some guidelines on how-to-do-it-yourself read Chapters 4, 5, 6, 7, 8, and 10.

Fourth, for an introduction to solar access planning read Chapter 9.

And finally for an overview of why passive solar design is a necessary and vital part of our energy future read Chapters 1, 2, 3 and 12.

b) How This Book Came To Be

The need for this book became increasingly clear to me in August while I was serving as a judge for the H.U.D. Passive House contest. Designs by-and-large wouldn't work or were overly complex, expensive, and relied too heavily on mechanical systems. Many designers were unaware of existing products and information which could have made their houses simpler, more effective, and less costly.

After writing **Village Homes' Solar House Designs** for Rodale Press in September and October I began work on the Passive Solar Catalog. The first letters went out to manufacturers and consultants on November 5. It has been an around-the-clock effort since then to get the catalog written and printed in time for the 3rd National Passive Solar Conference.

The response of manufacturers and consultants has been heartening. I only regret that the time constraints were so tight for this first edition. Many respondents simply could not meet the short deadlines. They will, however, be included in the next edition along with the many other consultants and manufacturers who have not yet been contacted.

The second edition will be available in January, 1980. It will feature a much broader range of products and more detailed product information, descriptions of consultant experience and expertise, and more technical articles. It will also include information on new developments in Passive Solar design and details on building performance. Retrofits and commercial projects will be featured in the section of examples. Contact us for details on advertising and availability.

Your comments on this catalog and suggestions for the next edition are encouraged. We are counting on you for information on new products and existing products we have not listed here. Descriptions of your experience with these products will also be appreciated. Photographs and descriptions of passive solar projects you have completed or undertaken would also be helpful.

The quality of our future may well depend on how quickly and how well you apply these simple principles to help reduce our dangerous dependence on foreign oil.

David Bainbridge
The Passive Solar Institute
P.O. 722
Davis, CA 95616
December 1978

WRITING: David Bainbridge
PHOTOGRAPHS: David Bainbridge, except where noted
TYPESETTING: Marc Fagan, Heather and Ann
ILLUSTRATIONS: Sue Rother, John Hofacre, and David Bainbridge
LAYOUT: David Bainbridge
PLATE PREPARATION: Johnny Print
PRINTING: Johnny Print

c) Acknowledgements

Completing a project of this magnitude in such a short time would not have been possible without the help of many people. Thanks to all who have helped, and particularly:

Bruce Behrens
Lola Brocksen
John Burton
Jeff's Camera
Marc Fagan
John Hofacre
Marshall Hunt
The Imagemaker
Johnny Print Copy Shop
Denny Long
Bruce Maeda
Don Marier
Bruce Melzer
Penny Nyland
Cindy Praul
Jeff Reiss
Steve's Place
Sue Rother
The Secretariat
Bob Schneider
Dave Springer
Virginia Thigpen

They have contributed much to the success of this effort. Any mistakes that remain are mine.

"Nothing would be done at all if a man waited till he could do it so well that no one could find fault with it."
Cardinal Newman

d) Dedication: To Loren "Tod" Neubauer, Professor Eneritus, U.C. Davis for his many years of research and support of passive solar design.

1950 THE SOLARANGER by Ralph Parks, **Agricultural Engineering**, p. 113 March Dr. Neubauer's solar modeling device.

1951 SOLAR HEATING AND THE SOLAR HOME 3 pages mimeo

HOUSE COOLING IN A WARM DRY CLIMATE Progress Report No. 2 Calif. Prof. 1536 USDA Proj. W-8 3 pages mimeo

1956 HOUSE COMFORT IN HOT WEATHER for Farm Structures Conference Jan. 27 3 pages mimeo

ENVIRONMENTAL INFLUENCE ON ORIENTATION AND HOUSE DESIGN TO IMPROVE LIVING COMFORT with G.J. Everson and R.B. Deering, **Journal of Home Economics** Vol. 46 No. 3 p. 161-167 March

1958 TEMPERATURE CONTROL FOR HOUSES with R.B. Deering and Virginia G. Kay **Journal of Home Economics** Vol. 50 No. 3 p. 175-184 March

1959 COOLING PRINCIPLES FOR BUILDINGS 7 pages mimeo

SUMMER HEAT CONTROL FOR SMALL HOMES with R.D. Cramer **Transactions ASAE** Vol. 2 No. 1 p. 102, 103, 105, 1959

1960 PRINCIPLES AND PARAMETERS OF SOLAR ENERGY COLLECTION AND USE with F.A. Brooks and C.F. Kelly, Presented 1960 Winter Meeting ASAE Dec. Memphis, Tenn., Paper No. 60-818 7 pages + cover

1961 THERMAL EFFECTS OF FLOOR CONSTRUCTION with R.D. Cramer, **ASHRAE Journal** Jan. 1961 p. 57-61, 72, 73 Presented ASHRAE Semiannual Meeting Chicago, Ill. Feb. 13-16

FARM BUILDING DESIGN with Harry B. Walker, Prentice-Hall, Englewood Cliffs N.J. 611 pages.
CHAPTER 10 - HEATING AND INSULATION
CHAPTER 17 - ENVIRONMENTAL DESIGN AND ORIENTATION

1964 TEMPERATURE CONTROL OF SOLAR RADIATION ON ROOF SURFACES with R.D. Cramer, Melvin Laraway, **Trans. ASAE** Vol. 7 No. 4 p. 532-434, 438 Presented ASAE Annual Meeting Miami, Florida, June 1963 Paper No. 63-403

1965 SHADING DEVICES TO LIMIT SOLAR HEAT GAIN BUT INCREASE COLD SKY RADIATION with R.D. Cramer, **Transaction ASAE** Vol. 8 No. 4 p. 470-472, 475 1965 Presented at Winter Meeting ASAE New Orleans, La., Dec. Paper No. 64-937

DIURNAL RADIANT EXCHANGE WITH THE SKY DOME with R.D. Cramer, **Solar Energy** Vol. IX No. 2 April-June p. 95-103

1966 SOLAR RADIATION CONTROL FOR SMALL EXPOSED HOUSES with R.D. Cramer, **Transactions ASAE** Vol. 9 No. 2 p. 194, 195, 197 1966, Presented at Winter Meeting ASAE Chicago, Ill., Dec. Paper No. 59-826

THERMAL EFFECTIVENESS OF SHAPE - I with R.D. Cramer, **Solar Energy** Vol. 10 No. 3 p. 141-149

1968 EFFECT OF SHAPE OF BUILDING ON INTERIOR AIR TEMPERATURE with R.D. Cramer, **Transactions ASAE** Vol. 11 No. 4 p. 537-539, Presented at Winter Meeting ASAE Chicago, Ill., Dec. 1966 Paper No. 66-913

EFFECT OF SIZE, SHAPE, COLOR AND ORIENTATION ON TEMPERATURE CHARACTERISTICS OF MODEL BUILDINGS, Presented Annual Meeting ASAE Logan, Utah, June 1968 Paper No. 68-413, 13 pages + cover

1969 SOUTHERNMATION: HEAT CONTROL FOR FARM BUILDINGS, Presented Annual Meeting ASAE, Phoenix, Arizona, March, Paper No. PCR-69-117, 15 pages + cover

TOWARD COOLER BUILDINGS, **Agricultural Engineering**, Feb. 1969 p. 80, 81 Condensation.

1970 OPTIMUM ALLEVATION OF SOLAR HEAT STRESS ON MODEL BUILDINGS, Presented Annual Meeting ASAE Minneapolis, Minn., July, Paper No. 70-401 15 pages + cover

1972 ORIENTATION AND INSULATION: MODEL VERSUS PROTOTYPE, **Transaction ASAE** Vol. 15 No. 4 p. 707-709 1972, Originally titled SOUTHERMATION AND INSULATION: MODEL VERSUS PROTOTYPES, Presented Annual Meeting ASAE Pullman, Wash., June 1971 Paper No. 71-407

SHAPES AND ORIENTATIONS OF HOUSES FOR NATURAL COOLING, **Transactions ASAE** Vol. 15 No. 1 p. 126-128 1972, Presented Annual Meeting ASAE Lafayette, Ind., June 1969, Paper No. 69-423

1974 **A STRATEGY FOR ENERGY CONSERVATION: PROPOSED ENERGY CONSERVATION AND SOLAR UTILIZATION ORDINANCE FOR THE CITY OF DAVIS, CALIF.**, with J. Hammond, M. Hunt, and R.D. Cramer, available from LIVING SYSTEMS, Route 1 Box 170, Winters CA, Price $5.00

1977 **DAVIS ENERGY CONSERVATION REPORT** with Jon Hammond, Bill Kopper, David Bainbridge, available from Living Systems, $7.50

1978 ANNUAL TEMPERATURE CONTROL by Passive Solar Design, with Gary Starrand, Bruce Melzer, Presented at the Annual Meeting ASAE, Logan, Utah, June 27-30

Dr. Neubauer has also published many papers on the strength of wood construction and adobe.

And the rest of pioneers:

F.A. Brooks
George Keck
Doug Balcomb
Peter and Florence van Dresser
Harold Hay
Victor and Aladar Olgyay
Steve Baer
Felix Trombe
Norman Saunders
Jon Hammond
Marshall Hunt
Jacques Michel
Bruce Anderson
Amory Lovins
Dave Wright
Baruch Givoni
Buck Rogers
Ken Kern
Virginia Thigpen
Karen Terry
John Hofacre
Mike and Judy Corbett
Bill Migenback
J. Palmer Boggs
Bill and Susan Yanda
Michael Riordan
Ed Mazria

And to my parents - who have always lived responsibly.

e) Table of Contents

1. What is Passive Solar Design
2. Why Passive Solar Makes Sense
3. Where Passive is Now
4. The Foundations of Passive Heating & Cooling
5. An Introduction to Passive Solar Heating & Cooling
6. Twelve Examples
7. Passive Solar Water Heating
8. Daylighting
9. Solar Access: Planning for Passive Solar Design
10. Build It Yourself
11. Access to Components and Consultants
12. Passive Solar and Energy Independence

Chapter 1. What Is Passive Solar Design

Many of the most effective and workable solar applications for new and retrofit applications are those called "Passive" Solar Systems. Although there are many definitions of this term several points are common to most or all of them. Passive solar systems are those that:

- Use energy from the sun and climate resources to provide heating and cooling.
- Do not rely on external sources of energy to function. (They will function even if the power goes off).
- Decrease the building's use of auxiliary energy.
- Combine collection and storage.
- Are simple and low in cost.
- Generally have a very long lifetime.
- Require little non-renewable energy to build and maintain.

The use of the term Passive Solar Design has been widely used and is now institutionalized. It is in many ways a poor choice because it sounds very "weak". The use of "direct" systems has been proposed to remedy this but has not caught on. Other writers have proposed the use of "Natural Systems" which is perhaps equally value laden and therefore inappropriate.

It is a just commentary on our society that Active systems, which won't work without fans, pumps, controls and power, are given the strong name and that Passive Systems which typically run by themselves, are given the weaker name. But then as Malcolm Wells puts it, "An SST, smashing its way through the sky by burning lakes of fuel is high technology. A bicycle, run by a carrot-eating maching, and thereby operating at a far, far lower cost per passenger-mile is low technology."

Three examples will perhaps more clearly illustrate what passive solar design is:

1. Passive Space Heating

A south window is the simplest passive heater. Any cat can find the warmest spot in the house on a cold winter day - in the sun streaming in a south window.

2. Passive Space Cooling

Any dog can find the coolest spot on a hot summer day - under a big shade tree with a gentle breeze. It combines ventilation, shade, and evaportation to provide comfort - much tike many passive solar systems.

3. Passive Solar Water Heater

A black water container sitting in the sun is a simple hot water heater. Putting it in an insulated box with south glass will make it work better.

And that is really all there is to it! The basics of passive solar design can be understood by anyone (even a cat or dog). Applying them to your house, office, or farm is straightforward, simple, and economical. You don't need to have a Ph.D in Engineering and an Architectural degree to be successful. (In fact, not having one sometimes seems to help!) So take a careful look through this book and take your first step toward energy self-sufficiency. It will be interesting, fun, and very rewarding.

A simple passive solar home

A passive solar retrofit benefits everyone

A passive solar water heater near completion

Chapter 2: Why Passive Solar Makes Sense

Now that you know what Passive Solar is let's take a look at why it deserves much greater attention than it has received to date. The following could and perhaps should be expanded and added to but I think these 15 reasons why passive solar makes sense will prove sufficient. Chapter 3 explains why passive is still largely unknown even with all of these sound reasons for favoring it over active solar or non-renewable fueled alternatives.

1. Heating and Cooling

Passive systems can provide both heating and cooling by using the sun and climate resources. This reduces the cost, energy demand, and increases the comfort levels.

2. Simplicity

Passive systems utilize basic principles, common materials, and construction skills that are available locally. This simplicity puts passive systems within reach of everyman, both here and abroad. The simple materials ensure that most passive systems will have very low "net" energy cost as well which is very important.

3. Reliability

Passive systems are inherently reliable as they have few or no moving parts. In addition, they will work even if the "power" goes off and can provide some security in an increasingly troubled world. The biggest advocates of solar greenhouses in Vancouver, B.C. are the older British couples who went through the blitz in England and know what it means when the power goes off.

4. Reduce First Cost to Builder (and Potentially for Buyer)

A well designed passive house will almost always have a lower first cost than a standard house with air conditioning and central heating. This saving can be used for greater profit or as a pass through to reduce the sales price.

5. Reduce Operating Cost for Buyer

An even bigger incentive is the potential for reducing utility bills for the buyer. This can be used to help the buyer get a bigger loan, because banks are beginning to consider expected utility bills in their evaluation of loans. Savings are particularly attractive for commercial applications, although they have received little attention.

6. Comfort

Passive buildings, contrary to common belief, are extremely comfortable. This is surprising to many, and must be experienced to be believed. The comfort results from more balanced surroundings, particularly the radiant temperatures and humidity. Further details are included in chapter 4 under Human Comfort.

7. Quiet

Passive structures are also very quiet due to mass, added insulation, and double glazing. This lack of noisy and irksome furnaces, compressors, fans, and ducts helps contribute to peace of mind.

8. Health

Passive solar heating and cooling systems are much healthier than regular mechanical systems for a variety of reasons. These include: lack of air pollutants associated with combustion; improved humidity regime; and better ion balance. This area has only recently been explored. Studies on indoor air pollution from existing stoves and furnaces are not encouraging. Levels often exceed E.P.A. long term exposure and as one researcher says, "It is thus essential that air pollution characterization of the indoor environment be undertaken in order to assess correctly the human exposures in epidemiological studies..."

9. Ease of Mass Marketing

Once the barriers mentioned in the next chapter are overcome, market penetration will be very rapid due to the simplicity, low cost, and readily available materials and skills needed for passive systems. While this is particularly true for new buildings, it will also prove true for retrofitting existing structures.

The ease of mass marketing is primarily a result of three factors. First, the materials used are commonly low in cost and readily available. Second, the skills and tools needed to manufacture most passive solar components are readily available and low in cost. And finally, passive solar systems can in most cases be installed by local labor with tools and skills already in use.

All three of these factors are important. The availability of materials in the local market insures that few supply hang-ups occur. Installation can occur rapidly without fear of delivery or manufacturing problems. The low materials cost reflects the low energy cost of most materials used and ensures that new businesses can get started and that even rapidly increasing energy prices will not disrupt the market.

The fact that skills and tools needed to manufacture passive solar components are readily available and low in cost is also very encouraging and suggests that very rapid market penetration can occur. For example, steel tanks for water storage in a passive solar house can be welded by almost any local welding shop. Thermal curtains and shutters, another vital component, can be manufactured by almost every drapery shop or seamstress, or cabinet and door company using standard techniques.

Finally, these passive components and passive solar systems can be installed by existing contractors and trades people with their existing tools. No special equipment is needed nor is sophisticated training required. This ensures rapid delivery, fast penetration of the market, and good profitability for the installers.

10. More Jobs

An exact estimate of the employment impact of widespread use of passive solar systems is relatively difficult to calculate. However, existing information does provide a rough comparison of employment impacts of a given investment in passive solar versus active solar or traditional sources used by public utilities. An investment of $100,000 will provide 10 jobs in passive solar, 5 jobs in active solar, or 1 job in a public utility.

This figure can be used to generate an estimate of the projected employment impact of investment in passive solar systems in the United States from 1980-1985 based on the estimate of desired investment in solar systems. Assuming the estimates made in the next chapter are reasonable (they are probably conservative) passive solar systems could provide 1,000,000 jobs per year between 1980-1985. Much of this work will be in retrofits and it will help balance out loss in construction jobs from higher interest rates.

11. Local Jobs

These jobs are primarily local jobs - and this is also very desirable. First, it can reverse local and national balance of payments problems. Money now going to distant heater and air conditioner manufacturers, and money flowing to Saudi Arabia for the oil to fire generating plants can now remain and be active in the local market place. The tremendous value of this decentralization should receive more emphasis and research than it has to date.

12. Regional Identity

One of the more subtle, positive impacts of widespread use of passive solar systems will be the re-establishment of regional architectural types. This diversity and identity have values which only recently have been realized. The universal house and office now applied indiscriminently in New York, California, London, Cairo, Delhi, and Toronto will be replaced by more comfortable, efficient, and appropriate designs. When we get off a plane in the future it may not be necessary to consult a map to find out where we are.

13. Peak Load Reduction

Although utilities have in most cases studiously ignored passive solar systems as a load management tool, (rightly so according to the corporation balance sheet which bases profit on total investment), they are increasingly interested in any methods which can help reduce peak power demands. Thanks to some courageous decisions by the California Energy Commission and the Public Utilities Commission we may see passive systems promoted by the utilities to help them out of their predicament. This will be to society's benefit as passive solar systems cost about 1/100 as much as adding generating capacity and will employ 6-10 times as many people for a given investment.

14. Environmental Impact

In addition to reducing peak load, passive solar systems have few detrimental environmental impacts. The materials used are by and large renewable, non-toxic, recyclable, and labor intensive. This is in contrast to virtually all of the other methods of generating energy, including active solar systems. When compared to the impact of coal, nuclear, or hydro power there are "no" adverse impacts.

15. Saving to Society

The reduced energy use made possible by the passive systems also provides several dollar savings to society. It reduces the need for investment in additional power plants - freeing capital for better uses. Consider that the expected capital requirements for large central power plants was calculated at $1 trillion between 1976 and 1985. That money would easily build 10 million new passive solar houses and remodel all of the houses in America in the same period. Which is the better investment?

The use of passive solar systems has been "passive" long enough. The need for action is inescapable. Many passive systems are well understood and many more are being developed. The sun is shining. Let's get to work.

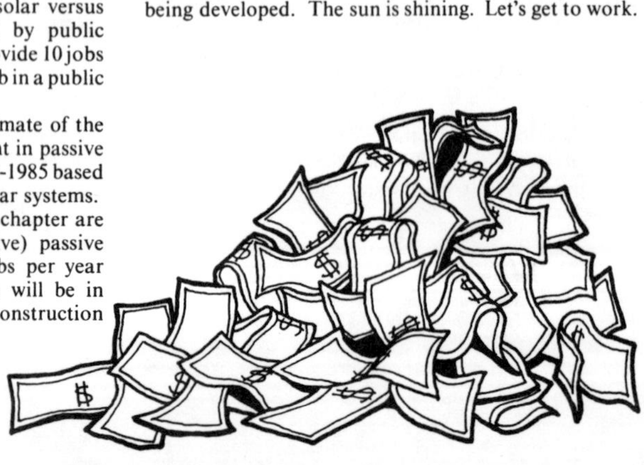

Chapter 3: Where Passive Is Now

1. Introduction

Passive systems have many advantages over the more traditional mechanical systems relying on nonrenewable energy sources. Yet despite these advantages passive systems have been studiously ignored in the many solar studies conducted by the utilities, "think" tanks, and various government agencies and passive solar is just that, passive. The reasons for this are not hard to identify and offer useful insights concerning what must be done both to promote passive solar systems and; hopefully, to reevaluate the method in which our society approaches problems. The following discussion identifies many of the obstacles to widespread use of passive systems.

The discussion of this somewhat gloomy topic is followed by a brief look at the long and successful history of passive solar utilization and an evaluation of the impact passive solar design could have in the near future - if a commitment to use passive solar is made now. The role of rapid passive development as a critical step toward a more energy independent U.S. is explored further in Chapter 12.

2. Obstacles to Passive Solar Systems

A. Ignorance

The most obvious obstacle passive solar development faces is widespread ignorance of existing information. With our present knowledge it is possible to build low-cost, effective, passive houses in most climates with good performance prediction. Yet few designers, architects, and engineers have a working knowledge of the vast body of literature now available. In fact, the general attitudes above passive solar systems are completely wrong and hold that "passive systems are expensive, uncomfortable much of the time, and require radical changes in lifestyle." The Passive Solar Institute was specifically established to overcome this obstacle.

B. Lack of Information

Although much is known about Passive Solar design and good buildings can be built using existing information, there are many gaps in our existing knowledge. The deficiencies should be clearly identified and research should be undertaken to improve our understanding of passive system operation. For example, the knowledge of heat transfer characteristics of floor slabs is based in large part on an imperfect study conducted in the 1950s. Before we can accurately predict the performance of a passive house (or any house for that matter) we need to have a better understanding of this simple problem. Far too much attention has been focused on very expensive computer models of performance when the basic subroutines and equations are imperfect. As the tried but true computer parable goes, "Garbage in, garbage out." It is a tragic fact that we know more about nuclear reactor behavior than we do about the thermal performance of a house. We need to support more basic research in this field.

C. Subsidies for Nonrenewable Energy Sources

An equally large problem is posed by the many tax and pricing subsidies for nonrenewable energy. If Federal programs, beginning with Rural Electrification and blossoming with the Bureau of Reclamation, T.V.A., and etc., had not favored non-renewable sources we might have had uninterrupted development of solar, wind, and other renewable energy sources.

However these subsidies have been made and have reached staggering amounts. Estimates in three recently completed studies of Battelle, Cornell University, and the Department of Energy suggest that the subsidy from rolled in pricing, tax credits, and average cost pricing may exceed 100 billion dollars per year for non-renewable fuels. On a cost per million BTU basis the most comprehensive study suggests the subsidies excluding rolled in pricing are $.09 for oil, $.03 for gas, $.015 for coal, and $6.50 to $7.00 for nuclear electricity.

These subsidies make non-renewable fuels more competitive with passive solar design that they would otherwise be. They should be dismantled and removed because they encourage misallocation of resources and favor increasingly dangerous dependence on foreign oil.

D. Special Interest Domination of Government Spending

An effective lobby will be necessary to fight for more balanced investment by the Federal and State Governments. A look at spending in Solar and Nuclear development to date illustrates the scope of the problem. While much of this money shouldn't be spent at all (we feel private enterprise in a freer market would allocate resources better), if it is going to be spent it should be spent on the right things, and passive solar is one of them. Retrofitting Federal and State office buildings with such things as shadescreen and insulated drapes would be an excellent market stimulus and example for the private building sector.

This special interest domination is also reflected in the size of companies receiving most of the research money. An unpublished study by the Office of Management and Budget credited small business with over 1/2 of the industrial innovations between 1953 and 1973. Yet, despite that impressive fact, and although just 16 small technology firms created 25,558 jobs in the same period, only 4% of Federal payouts went to small firms. An uncirculated memo which "fell through the cracks" between the Ford and Carter administrations stated it very clearly, "there is considerable evidence that the small proportion of federal research and development work that is being awarded to small technologically based firms is contributing to a serious loss of high technology capabilities in our nation."

The Federal Department of Energy has shown particular favor to the big, often inefficient, corporate bureaucracies with strong lobbies and weak research development and staffs. They have awarded only 2.6% of their research money to small firms. It would be interesting to see some of the progress that might have occured if this proportion had been reversed.

Although we would prefer to see much less Federal involvement in research and development, if the money is to be spent it should be spent wisely. The success of the recent Appropriate Technology grants program in the western U.S. suggests what might be done. Simple application forms, quick review, and relatively low grants (about $10,000 in most cases) helped support many useful and valuable research efforts.

E. Government Regulation

Although the impact of government regulation has been less adverse for passive solar than it has been for active solar systems it is still a problem. The numerous building codes and energy conservation regulations adopted by local and state governments are probably the worst offenders, but certain Federal regulations are also involved.

The need for a drastic overhaul of existing regulations is most readily apparent in the building code requirement that the heating system must be able to maintain an air temperature of 70 F three feet off the floor. This neglects both common sense and human comfort and should be dropped. In many cases it requires the installation of expensive gas or electric backup systems which will never be used.

Other cases have arisen recently where south glass is discouraged or outlawed, even if it is appropriately shaded, mass backed, and controlled with insulated drapes. California's Energy Conservation Code section is better than some, yet it requires complex steps to justify more than minimal glass. The process is so

The Path Not Taken

Total NonRenewable fuels subsidies 1920-1977 greater than 1 trillion

- Mt. Everest

- Oil subsidies *77.2 billion 1920-1976 (1976 dollar)

- Commercial Nuke* 24.4 billion (1975 dollar) 1960-1977

- Rolled in price subsidies for electricity in 1978, 20 billion

Rolled in price subsidy for oil in 1978, 14 billion

- Great Redwoods (400')

Total Solar, to 1977 400 million
Solar Heating and Cooling 100 million
Passive Solar 2 million

*Not including rolled in pricing subsidy

The 1978 budget for passive solar in California is around $200,000 while the Federal budget, although much larger, is still only about $12 million.

involved that there are few groups in California doing forms, often at exhorbitant prices. We need to do much more to encourage people to use passive solar systems. We should not discourage them.

F. Overspecialization

The overemphasis on hardware and high technology resulting from this type of investment has lead to overly specialized experts and disciplines which all too often miss the forest for the trees. Their narrow vision fosters bad design and poor solutions to complex problems. All too frequently, a specialist can only see his "solution" to the exclusion of others that are much more viable.

Architects and engineers in particular have become addicted to mechanical systems and energy transfusions to make designs work. Buildings are no longer considered as a total system which must fit the local climate but as a climate independent collection of component systems, each independent of the other, and each vastly oversized to ameliorate the loads caused by the oversizing of other systems and the failure to consider the local climate.

This practice would be bad enough if it was restricted to the more affluent Western countries, but it has, through our educational institutions, been transferred to scores of Third World architects and engineers who have returned with their Western education to help destroy their countries building heritage and energy supplies. Rather than learn from us, our society should learn from the idigenous architects and builders who over the centuries have refined climatically adapted, responsible design.

Broader training should receive greater emphasis, particularly in the fields of architecture, engineering, and solar design. Of particular concern is the virtual absence of courses on climate, microclimate, human comfort, and environmental impact in programs ostensibly set up to teach building design, space conditioning, solar technology, and planning.

G. Hardware Fixation

One spinoff of this overly specialized approach is a fixation with hardware, gadgets, and the technological fix. This mentality makes it difficult to encourage people to switch to passive space heating and cooling as they don't see how the house can stay comfortable without "hardware."

David A. Bainbridge

H. Bias Toward High Technology Versus Low Technology

This bias toward expensive and complex hardware is representative of a much larger problem that has diverted attention from passive solar systems to active solar systems and other more expensive energy alternatives such as nuclear power. To be successful in academia and socially prominent one cannot study simple problems or espouse simple solutions, for if everyone can understand them you are no longer an expert.

I propose Bainbridge's law as a statement of the problem: "If any problem can be studied or solved simply, easily, and cheaply or with high cost, complexity, and difficulty, the latter method will always be favored." Thus, while it is respectable to do complex, expensive studies of concentrating collector systems for solar power towers which require expensive materials, equipment and controls it is almost heresy to build and test breadbox hot water heaters, which are simple, inexpensive and can be built by anyone with modest building skills. Fortunately, there are some areas of passive solar design where very sophisticated and expensive research should be undertaken. These should satisfy the most highly skilled "technotwits" as Amory Lovins jokingly refers to himself.

I. Market Failure

These biases toward hardware and high technology have also contributed to the market failure in passive solar development. The manufacturers of active systems are easily identified and have constituted a fairly effective lobby. Prices are easily calculated, profits made (not so easily), and efforts to support active solar systems have been fairly successful.

In contrast, passive solar systems rely in large part on simple materials which are usually available from traditional suppliers, for whom they are, in most cases, a minor market. Also, there needn't be specialized solar installers for passive systems - so no lobby has developed there either. And finally, because the passive system cost for material and labor is integrated with the total structure cost and usually low in cost, it is very hard to isolate the passive system cost and develop revenue for a lobbying effort.

This needn't be the case, as window, drape, door, insulation, tank, masonry, and other industries do have a legitimate stake in passive solar development and could be organized to become an effective lobby. The Passive Solar Institute will devote much of its effort to developing this nascent Passive Solar Industry with education aimed at potential members and product development studies to encourage new manufacturers. This program will also help manufacturers market existing products. A spinoff from this program will hopefully be the development of an effective passive solar lobby.

J. Mystification of Solar

In fact, the only lobbiests for passive at the present time are the passive designers, builders, and homeowners. Unfortunately, they have in many cases fallen victim to the problems already listed. As a result passive solar design is viewed by many people as a mystic art, incomprehensible to the layman. As part of this, some passive houses have been made more expensive than necessary, partly to justify the high cost some of the passive designers charge for their "visions."

K. Bias Toward Architecture as Art

This is in part a carryover from architecture as practiced today. The biggest prizes are not for functional buildings that are comfortable, but for buildings that look "artful," particularly when photographed (with 2/3 scale furniture). No better (or worse) example could be found than the recent California State Office Building Competition, conducted by the State Architect's Office in California. The winner, despite very good contest rules and requirements, was probably the worst building in terms of function, practicality, and cost; but it looked like a "Solar" building. Rather than choosing simple and reliable passive solar techniques it included unbuilt, untested, and expensive active systems. Removing this bias on form rather than function should receive much greater attention than it does because it makes so many of our buildings uncomfortable energy wasters.

L. The Martyr Complex

Many of these factors have contributed to the last problem I will isolate, a martyr complex among passive designers. In part, this is probably a logical reaction to a society which so clearly rewards "stupidity" and has provided so many frustrations for the pioneers who followed common sense instead. Their strength in going against the grain and doing "passive" solar systems in spite of these pressures must be recognized and applauded; but their bullheadedness has become an obsession for some, making them hard people for less idealistic groups to deal with. This factor has played a part in retarding mass marketing and widespread adoption of passive systems.

Lest it seem that these problems are overwhelming, I will briefly review the history of passive solar and reiterate some very strong reasons why passive will become accepted and widely used in the near future and the impact it can have on energy use in the United States.

3. A Brief History of Passive Solar Development

Passive solar heating and cooling has been understood and practiced for thousands of years. Primitive architecture almost always uses passive heating and cooling to the maximum extent possible, given certain limits due to their available materials. Four examples will show how well developed passive heating and cooling were in ancient times:

A. MESOPOTAMIA 5000 B.C.

Although the roots of architecture reach back many thousands of years, some of the first masonry structures are from the Sumerian culture in Mesopotamia. These buildings, like the Pueblo's and cliff dwellings of Mesa Verda (1200 AD) used simple, readily available materials and good design to provide comfortable conditions in a harsh climate. Few windows, often with primitive shadescreen, were used to reduce heat gain and heavy walls stored the sun's heat for the cool desert nights. Wells were shaded to keep the water cool and the entrance courtyard wall on the north was used to provide a cool spot for comfort during the hot days. The following graph illustrates the thermal performance of a similar structure, the pueblo from the southwest. It is a better shelter than many houses and offices in "modern" America.

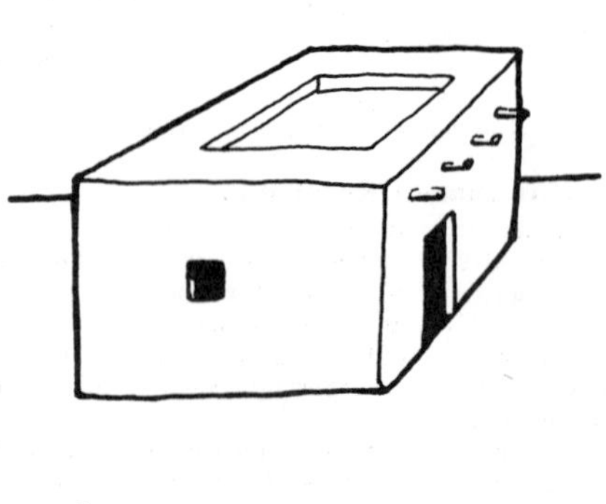

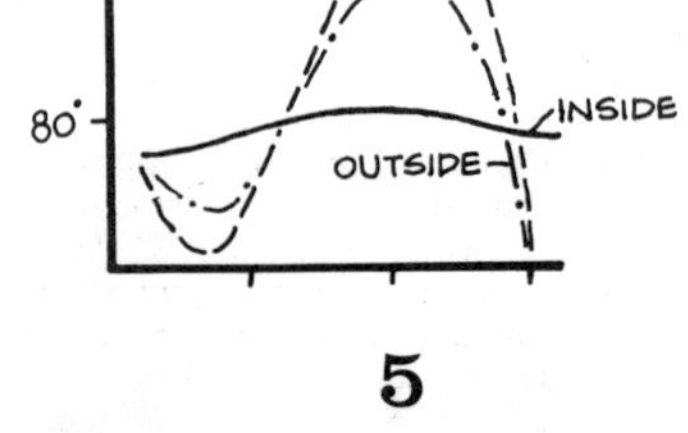

B. GREECE 3000 B.C.

The first written description of passive solar heating and cooling comes from over 3,000 years ago in Greece.

> *"In houses with a South aspect the sun's rays penetrate into the portico in winter, but in the summer the path of the sun is right over our heads and above the roof so there is shade. If then this is the best arrangement, we should build the south side loftier to get the winter sun . . ."*
>
> —Socrates

Greek city planning laid out streets on an east-west axis to maximize winter solar heating and minimize summer overheating.

C. ROME 100-500 A.D.

The Romans were the first civilization to make sheets of glass. They built buildings using the lessons learned in Greece with the added advantage of south glass rather than merely south openings. The famous Roman baths were solar heated with large south windows. Houses also used south glass effectively. In about 500 A.D. the violation of a person's solar access became a civil offense.

> *"If our designs are to be correct we must at the outset take notice of the countries and climates in which they are built. One style seems appropriate to build in Egypt, another in Spain, a different kind in Pontus, one still different in Rome, and so on with lands and countries of other characteristics. This is because one part of the earth is directly under the sun's course, another is far away from it, while another lies midway between these two. Hence, as the position of the heaven with regard to a given tract of the earth leads naturally to different characteristics, owing to the inclination of the circle of the Zodiac and the course of the sun, it is obvious that the designs for houses ought similarly to conform to the nature of the country and to the diversities of climate."*
>
> *"The rooms for the hot and tepid baths should be lighted from the southwest, or, if the nature of the situation prevents this, at all events from the south, because the set time for bathing is principally from midday to evening."*
>
> —Vitruvius

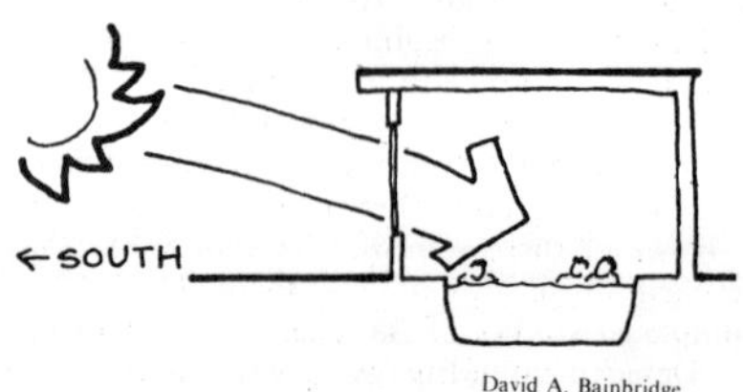

David A. Bainbridge

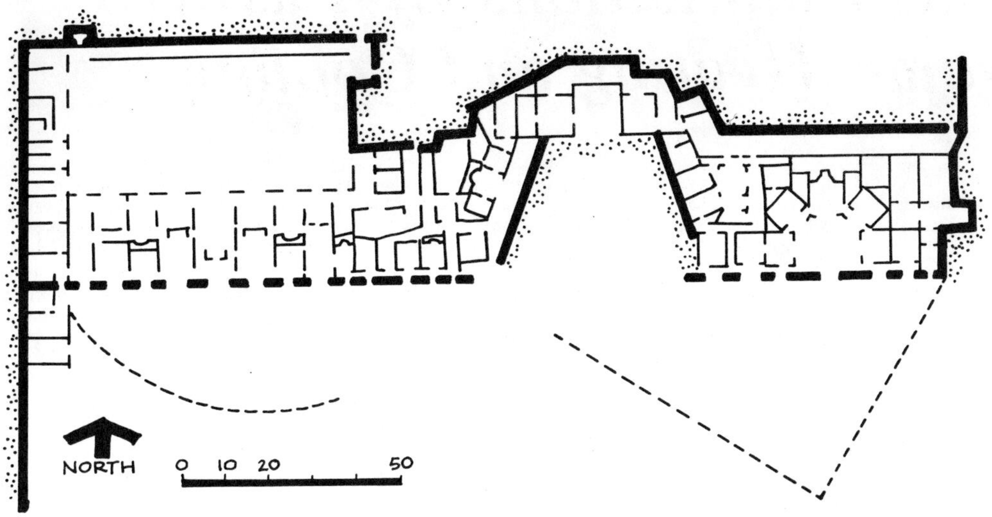

from: Roman Construction in Italy from Tiberius through the Flavians

Marion E. Blake
Carnegie Institute, 1959

D. PERSIA 900 A.D.

The final example comes from the area now known as Iran, and involves the very sophisticated use of the sun and climate resources to provide heating and cooling. These buildings not only utilize massive construction to store the sun's energy for use at night but include special features to provide increased cooling during the day. The diagram below shows one of the many systems in use. It includes a wind tower to catch the prevailing breeze, an underground tunnel to begin cooling the air with evaporation along the way and finally a fountain to further cool the air in the building - all passively. Many of these buildings are still operating well today - over 1000 years later. Will we be able to say the same for the Transamerica Tower?

1737 SOLAR GREENHOUSE

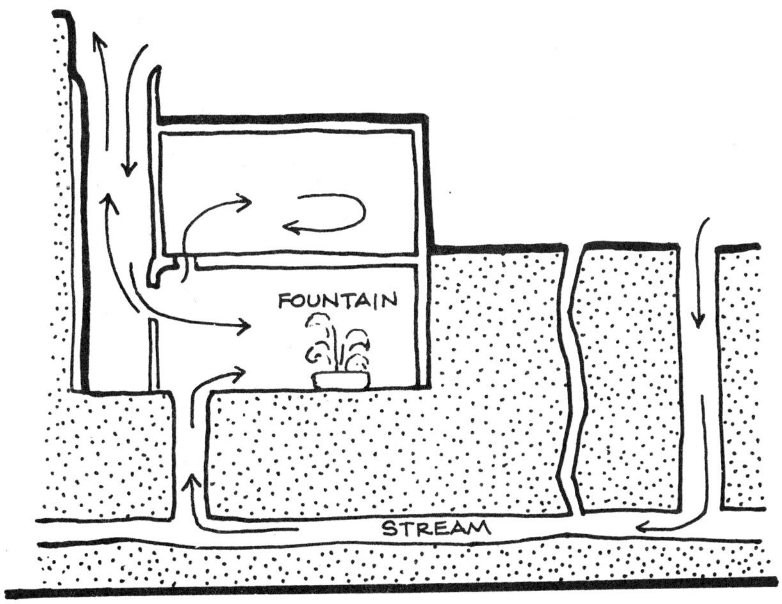

More recently many of these lessons have been ignored or forgotten. However, there are other highlights of passive solar design worth mentioning.

E. Selected Highlights of Passive Solar History

1600's Solar Greenhouse design in Germany and the Netherlands
1635 Sun slidrule, Oughtred
1737 **Landhuren, Lusthaven, Plantagian** published, P. dl. Court van de Voort, Netherlands.
1880 Solar pit greenhouse, Waltham, Mass.
1882 Mass wall solar heating, U.S. Patent, Morse.
1891 The "Climax" breadbox hot water heater, U.S. Patent and production.
1900 More than 1600 "Climax" heaters in use in California.
1903 **Solar Heat** published, C.H. Pope.
1936 **Use of Solar Energy for Heating Water in California**, published, F.A. Brooks
Solar Radiation as Related to Summer Cooling and Winter Heating, published H.N. Wright.
1940's Solar Subdivisions built near Chicago, George Keck
1945 Solar House Studies at Purdue, F.W. Hutchinson
1947 M.I.T. Solar House II, waterwall solar house.
1947 **Your Solar House** published
1948 Sunnycrest, simple passive solar house, Mr. and Mrs. Wendell Thomas
1948 **Climate and House Design: Physiological Study**, published J.W. Drysdale.
1950 Miller House, simple passive, Palmer Boggs and Mr. & Mrs. David Miller.
1950 The Solaranger developed, Loren Neubauer.
1953 **Climate and Architecture**, published, J.E. Aronin
1954 **Application of Climatic Data to House Design**, Published, Victor and Aladar Olgyay and Thomas Malone.
1955 Natural Cooling and Heating research, Loren Neubauer
1956 First test module, Odeillo, Felix Trombe and Jacques Michel.
1957 **Solar Control and Shading Devices** published, Victor and Aladar Olgyay
1960 Saunders House, simple passive, Norman Saunders.
1963 **Design with Climate**, published, Victor Olgyay
1967 St. Georges School, simple passive, Emslie A. Morgan.
1967 First mass wall house, Odeillo, Felix Trombe, Jacques Michel
1967 Phoenix roof pond test module, Harold Hay.
1969 **Man, Climate, and Architecture**, published, Baruch Givoni.
1972 Cherry Hill Office, underground, Malcolm Wells.
1973 Atascadero roofpond House, Harold Hay, Ken Haggard, Phil Niles Hammond House, waterwall retrofit, Jon Hammond.
1974 Winters roof pond house, Jon Hammond. Wright House, simple passive, Dave Wright. Dasburg House, simple passive, William Lumpkin.
1975 Solar Greenhouse project, Bill Yanda. Guadalupe project, Dan Areillo and John Yellot. Longview school, water wall, Jon Hammond, Dick Bertaurex. Village Homes started, Mike and Judy Corbett, John Hofacre. Davis Energy Conservation Building Code passed, mandate passive solar, Jon Hammond, Marshall Hunt, Dick Cramer, Loren Neubauer. Kelbaugh House, mass wall, Doug Kelbaugh. Tyrrel House, mass wall, Bruce Anderson. Krushke House, solar greenhouse, David Krushke.
1976 First Passive Solar Conference. University of Minnesota Library, Myrers and Tom Bligh. Benedictive Monastery, Steve Baer. **The Solar Home Book**, published Bruce Anderson Prince Edward Ark, New Alchemy Institute. Hunt-Thigpen house, water wall, Greg Acker and Marshall Hunt.
1977 Second Passive Solar Conference Bainbridge Water tank house, Virginia Thigpen **Homegrown Sundwellings**, published, Peter van Dresser. Smith House, solar greenhouse, Lee Porter Butler, John Hofacre 55% Solar Tax Credit in California, with passive eligible.
1978 First H.U.D. passive solar house competition, 600 entries. Federal solar tax credit passed with passive systems eligible. **The Solar Greenhouse Book**, published James C. McCullagh, editor. **Solar Heating**, published, Holly Antolini, editor. **Earth Sheltered Housing Design**, published, Underground Space Center, University of Minnesota. Feller House, warm wall, Paul Feller. Passive Solar Institute founded.

4. Potential Passive Utilization In The United States

The market penetration of passive solar heating and cooling, water heating, and daylighting will be very fast when the current ignorance of builders, buyers, and decision-makers is overcome. Passive systems will be accepted quickly because they have many attractive features and make economic sense.

All future housing built in the U.S. should include basic passive solar orientation. In addition, future planning should be oriented to provide for maximum ease of installation of simple passive systems with good solar access protection. This passive orientation can provide an energy saving of one third or more for space heating and cooling in all new houses. If a concerted effort is made four million of the units to be built by 1985 could be built with passive solar features, saving 400 billion BTUs a year in 1985.

In addition, with combined private and public investment (the public investment is needed to offset public subsidies of non-renewable fuels) of less than $12 billion dollars/year [less than 20% of what is now spent on nonrenewable fuel subsidies ($100 billion), advertising ($77 billion), or defense ($118 billion)] over half of the U.S. housing stock could be retrofit by 1985 with an annual saving of over 3 quadrillion BTUs a year.

This may seem impossible, but it is in fact attainable and may indeed be essential for the survival of the United States as a free and independent nation. Passive solar systems can contribute more to the national defense than military spending ever will.

For a more detailed discussion of this subject turn to Chapter 12.

David A. Bainbridge

Chapter 4: The Foundations of Passive Solar Heating and Cooling

1. Introduction

The first chapters described Passive Solar design and its benefits. This chapter will provide the foundation necessary to design a passive building. It is fairly lengthy and at times complicated so you may wish to skim over it and go on to the rest of the book. Return to this chapter when you need more specific information.

This chapter includes an introduction to the principles and methods involved in passive heating and cooling: heat transfer, heat storage, climate and microclimate, human comfort, and the effect of human activity on heating and cooling in buildings. Further reading on these topics is recommended for those who wish to gain a more comprehensive understanding of the foundations of passive solar design.

2. Heat Transfer

The design of passive solar systems requires a clear understanding of the various methods of heat transfer: radiation, convection, conduction, and evaporation. These are briefly discussed in the following sections.

A. Radiation

Radiation is the dominant heat transfer mechanism employed in most passive solar designs. Any object at a temperature greater than absolute zero (-273°C) **emits** radiation with the wavelength and intensity dependent on temperature. Radiation will always flow from a warmer object to a cooler object. This would ultimately result in thermal equilibrium if no other energy were added or taken away from the objects.

The graph below shows the ideal curves of radiation emitted from the sun and earth. Both are important in many passive designs. Shortwave radiation from the sun is collected for heating. Longwave radiation to space may be used for cooling.

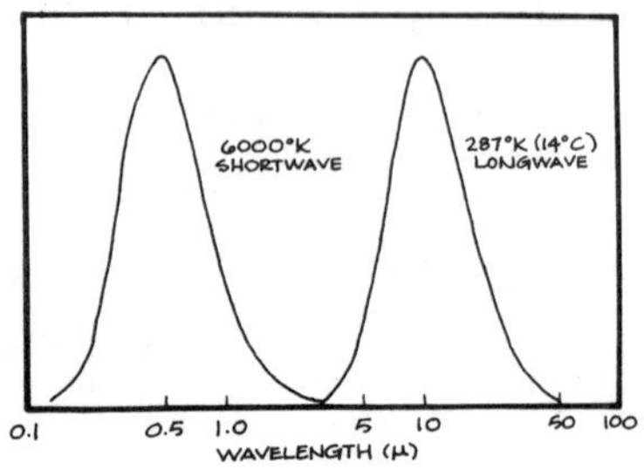

Building materials such as plaster, concrete and aluminum differ in their ability to emit radiation. The emittance may also vary with wavelength. The emittance of materials is important in many designs - particularly those that rely on both short and long wave radiation for heating and cooling. Those that collect energy effectively yet emit little energy, are known as selective surfaces. The table below presents the emittance for various materials that may be used in a passive or active solar system:

	Emittance
Rough concrete	97%
White plaster	91%
White enamel on galv. iron	90%
Concrete	88%
Black paint on aluminum	88%
Granite	41%
Galvinized iron (oxidized)	28%
Galvinized iron (new)	13%
Nickel black on galvinized iron	12%
Cupric oxide on sheet aluminum	11%

When radiation strikes a surface, it can be **reflected**, **absorbed**, or **transmitted**. The properties of various materials may be quite different and materials must be chosen to ensure that desired performance is achieved.

The angle at which direct solar radiation strikes a surface is known as the angle of incidence. The angle of incidence must be known to calculate the absorption, reflection, and transmission of solar energy received on a surface.

Reflection occurs when radiation bounces off a surface. This can be either specular or diffuse reflection. A mirror exhibits specular reflection while white paint exhibits diffuse reflection.

The term reflectance is used to describe the ability of a given surface to reflect radiation. The reflectance of a surface is generally given as a ratio of the reflected to the incident radiation, expressed as a percentage. This varies with wavelength and should be prefaced with the wavelength for which it has been determined. The reflectance can be described for both specular and diffuse radiation.

The following table lists the diffuse reflectance of various surfaces in visible wavelengths.

Surface	Diffuse Reflection as a Percentage of Total Radiation
Fresh snow	75-95
White paint	80
Old snow	40-70
Dry sand	35-45
Red brick	45
Granite	45
Concrete	40
Grass	25
Meadows/fields	12-30
Water (40° sun angle)	10
Dark soil	7-10
Tar paper	7

Shiny surfaces such as aluminum foil and water can demonstrate specular reflectance. Specular reflection can also be relatively important on rough surfaces at low angles. The following table illustrates reflectivity for surfaces with high specular components, again for visible wavelength.

Surface	Reflection as a Percent of Total Solar Radiation
Aluminum Foil	85
Water (surface in movement) 10° solar angle	50
Water (surface in movement) 2° solar angle	90
Dry sand 10° solar angle	80
Wet sand 10° solar angle	60

The radiation that is not reflected is either absorbed or transmitted. If it stays in the materials it is **absorbed**. For example a blackhose left in the sun heats up as it absorbs the energy from the sun. The ease with which it collects energy is referenced to as its **absorptance**. As with reflection, absorptance varies with wavelengths.

The table below presents absorptance for various materials in sunlight.

Absorptance

Flat black paint - 96%
Black tar paper - 93%
Galvanized steel - 65%
Concrete - 60%
Brick (red) - 55%

If radiation passes through the material, it is said to be **transmitted**. The **transmittance** of materials also varies with wavelength. It is necessary to know the spectral transmission of glazing for passive systems where the glazing may need to be transparent for both solar radiation and thermal radiation. Typical spectral transmission curves are shown below for common glazing materials:

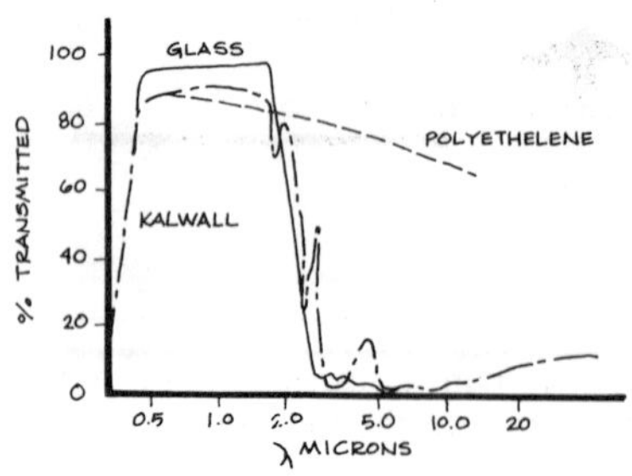

The manner in which the radiation is transmitted depends also on the index of refraction of the material. The index of refraction determines how much radiation will pass through a material and how much it will be redirected or refracted. A prism is a common example of refraction. It separates the different wavelengths and gives the familiar rainbow. Glass is typically better at transmitting radiation at acute angles than plastic. However, the ease of molding and shaping plastic into complex shapes can offset this deficiency.

Solar radiation is the dominant factor in almost all passive solar designs. To use solar energy effectively, we must understand the sun's path and the nature of solar radiation. The sun's position is described by its height above the horizon, altitude, and its bearing from the true north, azimuth. The change in elevation and azimuth over the season is a critical aspect of passive design. At 40° north latitude the sun is at an altitude of 30° at noon on December 21 and almost 70° during the summer. In summer it traverses an arc of 240° from East to West while in winter it only covers 120 degrees.

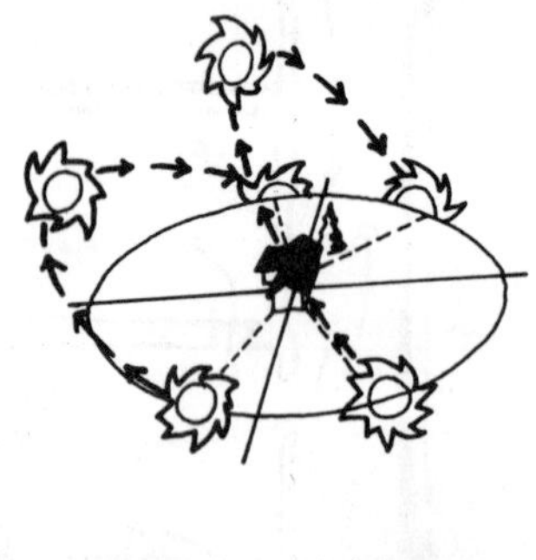

Before we go on, it may be helpful to describe solar radiation in greater detail. Although the sun is an efficient radiator, not all of the energy striking the edge of the earth's atmosphere reaches the surface of the earth. The reduction in energy is primarily due to reflection and absorption by carbon dioxide and water vapor in the atmosphere. The actual amount of direct radiation received varies with the atmospheric content, cloudiness, and the solar angles, which determine the path length. On the average 31% of the extraterrestrial solar radiation reaches the earths surface as **direct** radiation.

The radiation that reaches the earth after reflection or refraction is known as **diffuse** radiation. The amount will vary with the atmospheric content, cloudiness, and solar angles. On the average about 22% of the extraterrestrial solar radiation reaches the earth as diffuse radiation.

On a cloudy day diffuse radiation may account for almost all of the energy received at the surface. It is often assumed to be uniformly distributed over the sky for simplicity but the distribution is in fact far from uniform and the radiation is usually much stronger near the sun. The total of the diffuse and direct components of solar radiation reaching the ground is known as the **Global Radiation**.

The solar radiation received can be increased by using either natural or artificial reflectors. The amount of radiation received on a south facing vertical surface from a horizontal reflector can be estimated using the following formula and the table of reflectance given earlier.

Reflected Radiation = ½ (reflectance) (Global Radiation).

Adding the global and reflected radiation gives us the **Total Solar Radiation**.

This is particularly important for passive solar systems in the higher latitudes where a snowy surface to the south wall collector may add 50% to the total solar radiation in the winter.

Careful use of the different positions of the sun during winter and summer is critical. Houses can be oriented to exclude the sun during the hot summer months yet still admit the sun when it is needed for heating in the winter. The impact of shape and orientation have been thoroughly studied and the results indicate that the most desirable shape is rectangular, with the east-west length at least twice as long as the north-south width and most windows on the south and few on the east, west, or north. A two story structure of this type will perform better than a similar one story structure. These buildings work well because they minimize the critical east and west surface areas which can cause overheating in summer. This orientation reduces summer heat gain through both the windows and walls yet is excellent for collecting the low winter sun.

This change in sun path is also critical for cooling because the windows are the most vulnerable point for heating. Passive solar designs must exclude direct sunlight and minimize the impact of diffuse sunlight and reflected sunlight to stay cool in the summer.

Overhangs are very effective for solar control because they block the high summer sun. Extending the south overhang of the roof is not always cheap, but it is effective for summer cooling, permanent, and provides other benefits, including better weather protection for the south wall and windows.

This simple approach to overhangs does not work well in areas with a cool spring and hot fall. In those areas it would be desirable to have the sun shine in during the spring yet keep it out in the fall. The diagram below shows a moveable overhang, which is one solution to this problem.

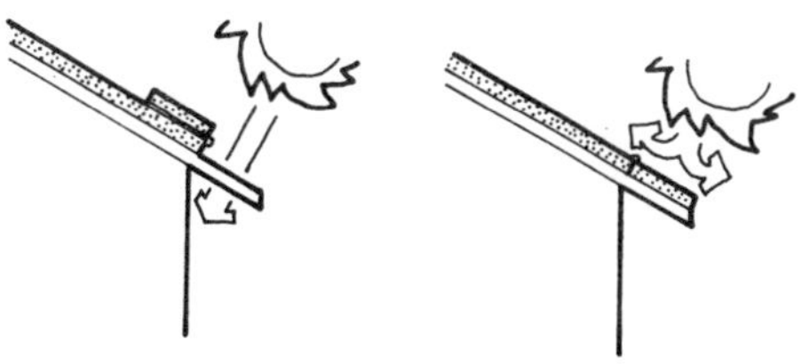

The unwanted morning and evening sun in summer can reach a window that is protected by an overhang during midday. Wing walls are an easy approach to prevent unwanted solar gain.

Exterior shades are more effective at keeping buildings cool than interior curtains or drapes because they block the heat from entering the building. Once the sun enters through the window the heat has already entered the house and although its impact can be minimized it is difficult to do so. Standard curtains and drapes may actually increase heat gain through an otherwise unshaded window, but an external shade will always help.

The simplest and least expensive method for exterior shading is a roll-down canvas awning or bamboo shade outside the window. This is particularly important on the east and west windows. A space for venting should be left at the top of the shade to prevent build-up of a hot pocket of air in the overhang. This provides considerable protection at a low cost. When the shade is not needed it can be rolled up or taken down. Where wind is a problem it is necessary to provide a ground anchor, preferably with a shock cord to prevent flapping.

Louvered shade screen made with bronze or aluminum is a more expensive method of exterior shading but it is also more durable. It consists of many tiny louvers that also serve as a bug screen. It can be used in standard screen frames that snap onto the windows.

More recently several companies have started manufacturing shade screens that use flat webbing instead of louvers. This material is less durable but it is also much cheaper. It may be a better choice in many cases, particularly where diffuse radiation or reflected radiation is a problem.

Curtains can increase heat gain unless they are designed properly. If they are well designed, they can provide solar control as well as reducing conducted heat gain. Drapes and interior shutters made specifically for solar control share some or all of the following features: highly reflective outside materials, high resistance to heat flow (R value of 5 or so), and tightly sealed edges at top, bottom and sides.

A typical drape specifically made for effective solar control is shown in the following drawing as an example of what to look for. See Chapter 11 for supplies and manufacturers.

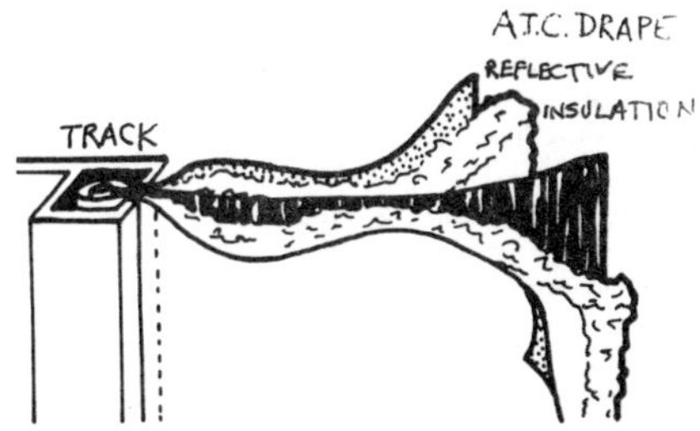

The final method of solar control for windows is the use of reflective films and coverings. Although these are widely advertised for window treatment they are usually the least desirable solution for solar control. The most obvious drawback is that they are typically permanent and cannot be removed in the winter when the sun is needed for heating.

Although insulated ceilings significantly reduce the heat gain from a hot roof it is often desirable to further reduce heat gain by using very light colored roofing. This can reduce roof temperature 50°F or more, which will help cool the house and also extend the life of the roof.

The next step is the use of materials which provide a cooler roof. Shakes stay particularly cool, followed closely by tile and shingles, with built-up roofs, galvinized iron, asphalt, and asphalt roll roofing trailing behind.

Other techniques can also be used to reduce roof temperature. The double roof used in the tropics and on ice houses in the north may be desirable to further reduce heat gain in extremely hot areas. The second roof should be light in color and vented thoroughly. Additionally, this double roof can be used for winter heating and nighttime cooling in the summer, if it is detailed properly.

A sprinkler system can also be used to reduce roof temperature. This may reduce roof temperature 5-10°F. In addition sprinkling may be used to augment night sky cooling effects. (See Evaporative Cooling.)

A sodroof may also be installed. The sod acts as insulation and adds evaporative cooling and thermal mass.

One of the most cost effective methods of solar control is simply good landscaping. Proper placement of trees, arbors with vines, and vines can provide effective and attractive solar control.

Full shade from vegetation can reduce heat gain through windows, walls and roof all at once. In addition, if deciduous trees are used properly they will not block the winter sun. However, it is important to keep trees primarily on the east and west rather than directly to the south because even the bare branches of a deciduous tree may block up to 45% of the winter sun.

Landscaping for shading is desirable because it automatically adjusts for cool springs and hot falls. Leaves may be just coming out in March (particularly if it is a cold year), but will still provide full shade in September (particularly if it is a hot year).

Freestanding walls and fences may also be used for exterior shading. These can be particularly effective on east and west windows and walls. These should be louvered or built with slats to reduce adverse wind effects. They will also prevent sideyards or courtyards from becoming heat traps.

Proper landscaping can reduce unwanted heat gain from reflection and diffuse radiation. Ground covers can be chosen to maintain cool surfaces and a better microclimate for living.

A typical house plan might include: tall trees for roof shade, deciduous vines on an arbor to the south to shade the windows, a dark ground cover to the south to reduce reflection, hedges to the east and west to shield the east and west walls and windows, and shrubs or small trees to the northeast and northwest to block late evening and early morning sun in the hottest period of the summer.

B. Thermal Radiation

Radiation at wavelengths longer than 3 microns will be referred to as **thermal radiation**. This is an important element of passive solar design as it covers much of the energy exchange within a room and the use of the cool night sky and cool north sky for radiant cooling.

The latter is often ignored or misunderstood. In addition to radiation received directly or indirectly from the sun, the earth also receives radiation from the water and gas molecules in the atmosphere. The difference between that received, and that reradiated to space is referred to as the net outgoing radiation when more is radiated than received. This radiation is primarily at wavelengths between 6-15 microns.

Most of the net outgoing radiation will occur to the cold night sky. The greatest radiation loss occurs directly overhead. Research has been done on the variation in radiation across the cold night sky. F.A. Brooks compiled a table of variation in net outgoing radiation with changes in zenith angle at Blue Hill, Massachusetts. He found that the rate was high down to about 20°.

Trees, walls, or structures can decrease night sky cooling. The reduction in radiative loss depends on the placement and also on the nature of the configurations of obstruction.

Net outgoing radiation can also occur during the day from a surface shaded from the sun and facing north. This is made possible by the cool spot in the sky that occurs at the point opposite the sun at right angles to the sun. Averaged over the day, the coolest spot is due north at an angle of elevation from the horizontal equal to ninety degrees minus the altitude of the sun at noon.

Clouds can greatly reduce night sky radiation as they increase the thermal radiation to the earth from the atmosphere. The exact effect depends on the height of the clouds, their density, and temperature. Relatively little work has been done to provide tables for calculating the impact of different cloud types on outgoing radiation. The table below presents experimentally determined values for the reduction caused by different types of clouds.

Percent Reduction for Different Cloud Types

Cloud Type	Percent	Elevation in Feet
Cirrus	16%	36,000
Cirrostratus	32%	24,000
Altocumulus	66%	12,000
Altostratus	80%	6,000
Stratocumulus	88%	3,000
Stratus	96%	1,200
Nimbostratus	99%	270
Fog	100%	0

C. Convection

Second in importance to radiation is **convection** or the transfer of energy through air flow. This may be flow caused by local differences in density of air (cooler air is heavier) or weather patterns with characteristic flow. The convective flow of air in and out of a building is known as infiltration. It is affected by design and placement of vents, windows, and interior walls and the operation of the house by the tenants. A well insulated house may lose almost half of its heat through infiltration. A fireplace may act as an air pump pulling cold air in through the cracks and exhausting warm air up the chimney.

A greenhouse stays warm primarily because the glazing prevents convection cooling—not by trapping radiation as is commonly thought. If the "greenhouse effect" of radiation trapping was a real phenomenon then polyethelene greenhouses would not work. The greenhouse demonstrates the importance of reducing convective heat loss.

A good measure of infiltration heat loss is the number of times the air in the house is replaced every hour. A typical American house may have 1.5 air changes per hour. Using careful caulking, sealing, and window design the Swedish have reduced infiltration to 0.2 air changes per hour. A heat exchanger air intake for ventilation may be desirable when infiltration is this low; see Chapter 11.

In the summer, however, we may wish to make use of convective cooling. One of the most economical and common methods of cooling uses ventilation with low temperature night or early morning air. Building design should incorporate full solar control and good ventilation to make the best use of convective cooling.

Convective cooling results from either natural winds, the density difference between warm and cooler air, or mechanical ventilation.

The simplest and most common form of convective cooling will be cross ventilation using natural breezes through operable windows or vents. Good ventilation with natural breezes requires careful design and orientation.

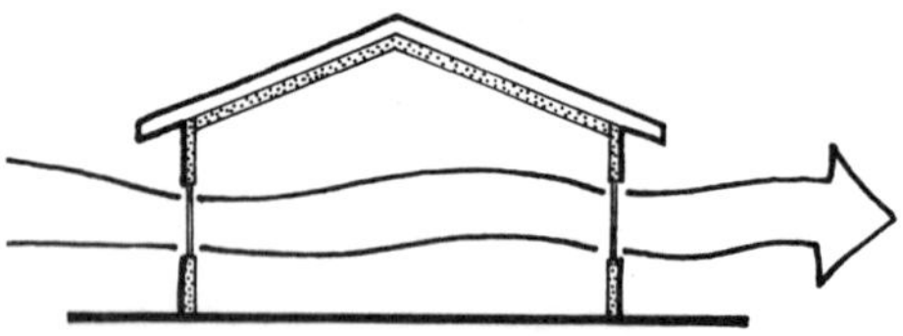

An obstacle like a building in the path of the wind will cause a high pressure area (+) to develop on the windward side and a low pressure area (-) to develop on the leeward side.

If there is only one opening into a room, air generally will not flow into it. This is known as a cave type structure. Tests have shown that ventilation in this type of room (which is most common in apartments) may be less than half that of a cross ventilated room for a similar area of window opening.

There must be an inlet and an outlet for air to flow well. These can be either windows, vents, or screened or louvered doors. For better air flow, the opening on the windward wall should be smaller than the opening on the leeward wall. A wind from an angle will also provide good ventilation.

The pressure difference between two points on the building, the vent placement and interior arrangement influence the potential for ventilation. The actual rate can be determined once the house is built to provide very accurate estimates of ventilation. In most design we simply try to provide for a very flexible ventilation and allow the occupant to adjust for comfort.

When using these principles of wind flow in building design, make sure that the air flow contacts the human body to create a cooling effect. Thus vents in the bedroom may be very low, while in the dining room a higher window would suffice. The type of opening and placement also influence air flow.

Placement of interior walls, halls and doors is also very important. Louvered doors and windows can provide cross ventilation while still providing needed privacy. Grills or louvered vents can be used to provide ventilation without a loss in privacy or security. The old metal grills in Spain and Mexico served this purpose.

We can use the wind even when it does not blow in a direction compatible with good solar orientation only if we use very good design. Exterior landscaping or a fence can be very effective in channeling wind into the house. This can also be accomplished by using casement windows as wind scoops. Operable wind flaps could also be used.

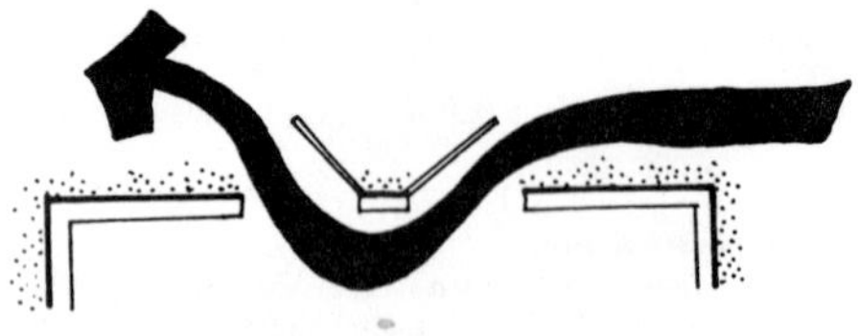

In areas where the wind almost always flows from the same direction a wind scoop may be incorporated in the design of the house or placement of house and garage. Where wind direction is more variable a roof vent that is effective with winds from any direction can be used.

Streets and landscaping should be laid out to protect natural breezes for cooling. The wind flow around windmills in Holland is protected by "wind rights" and a similar concept may be useful to consider in the layout of subdivisions and cities. Wind flow can be protected by running the streets at only a small angle to the wind, and using trees and landscaping to channel the wind into the houses. Where strong winds occur regularly landscaping and design can be used to block the unwanted gusts, yet still protect the cooling breezes.

When strong winds are common when cooling is needed careful detailing can reduce the impact of the wind yet still provide cooling. The sill vent is one method that can be used in windy areas. It will allow good ventilation without having drapes and other objects blowing around. It also provides protection from driving rains and allows an uninterrupted view through the glass without the visual interference of louvers.

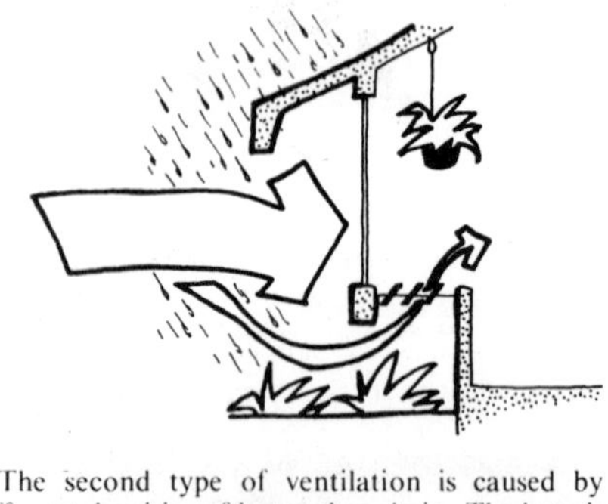

The second type of ventilation is caused by the different densities of hot and cool air. The hot air will rise and can be used to generate a cooling breeze. To get good ventilation this way it is desirable to have vents both up high and down low; see the Reiss house example.

The cooling can be increased by drawing from a pocket of cool air from under a suspended floor, or by drawing air through pipe buried in the cool ground. A hillside can provide an excellent source of cool air with a buried pipe network. As the air is cooled it will sink through a pipe into the house. Venting the attic is also helpful. A wide variety of vents are available, including gable vents, ridge vents, skylight vents, and different types of roof vents.

David A. Bainbridge

Added ventilation may also be provided by increasing this thermal stack even more. A sun chimney can be added to the roof. This can be either a pipe painted black or a more sophisticated solar heater with glass face and black interior. For enhanced ventilation into the evening, this may include thermal mass. For maximum flow the chimney should be very smooth.

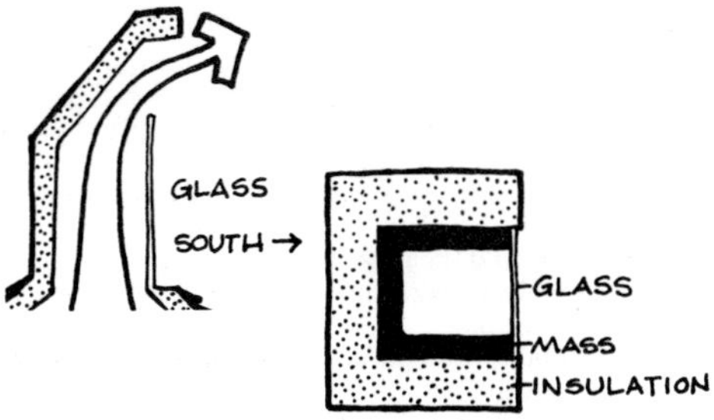

Both wind and density flow are usually combined and can provide all of the needed cooling in many areas. The wind tower from Iran is an example using both of these factors. Although it would be most applicable in commercial buildings it could also be included in a residential structure. The potential for convective cooling can be estimated from the high and low temperatures, wind patterns, and the diurnal temperature patterns. A wind rose is desirable for proper placement of vents, landscaping, and detailing. Monthly wind roses are desirable for your site.

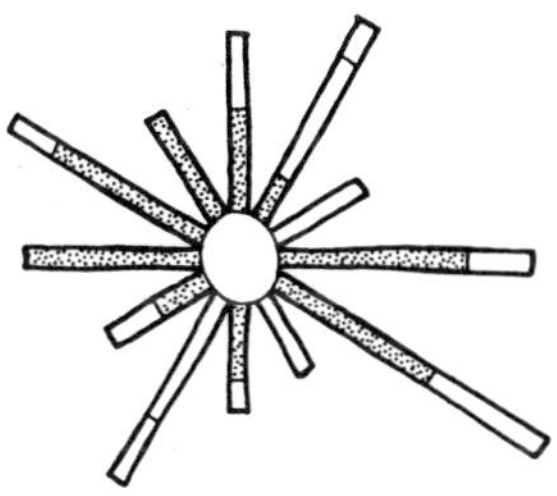

The potential rate of cross ventilation cooling can be calculated crudely using the following formula:

$$Q = 3150\ AV$$

where

Q = rate of airflow, cu. ft./hr.
A = area of inlets, sq. ft.
V = wind velocity, mph

This expression requires adjustment in cases where the area of outlets are appreciably different from the area of inlets as:

Area of Outlets: Area of Inlets	Value to be substituted for 3150 in above expression:
1:1	3150
2:1	4000
3:1	4250
4:1	5350
5:1	4400
3:4	2700
1:2	2000
1:4	1000

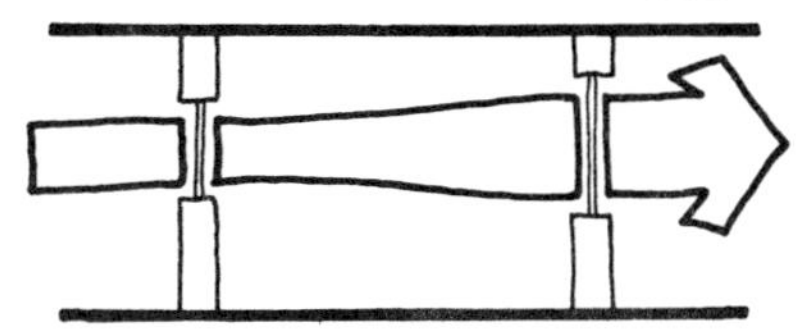

The value of the stack effect (for example on a clerestory house with operable windows) can be estimated from the following forula:

$$Q = 540\ A\sqrt{H(ti - to)}$$

where

Q = rate of airflow, cu.ft./hr.
A = area of inlets, sq.ft.
H = height between inlet & outlet, ft.
ti = avg. temp. at height H, °F
to = temp. of outdoor air, °F

This expression requires adjustment in cases when the area of outlets is appreciably different from the area of inlets according to the following ratios:

Area of Outlets: Area of Inlets	Value to be substituted for 540 in above expression:
5	745
4	740
3	720
2	680
1	540
¾	455
½	340
¼	185

In many areas, natural ventilation from wind and density flow will be all that is required. Other areas may require forced ventilation using fans and ventilators. A relatively small fan will usually suffice. An attic fan drawing from the house is often best. The size of the attic fan required for night air cooling depends in part on (1) the floor area of the house, and (2) the number of air changes desired. Where the night air temperatures do not drop much below 75°F, a large amount of air must be circulated.

A Casablanca fan may be used in place of an attic fan in some areas. These "Bogart" type fans use very little energy yet provide very good ventilation.

While a window fan is easier to install than an attic fan the former is limited in the area it can cool; furthermore, it is likely to be noisy. If installed in a hallway, a window fan can serve more than one room. If the fan is placed in a room, the doors to that room will have to be left open, and all windows adjoining the one in which the fan is located must be kept closed in order to prevent "short circuiting" of the air.

D. Conduction

The flow of heat from molecule to molecule is called **conduction** and can be a very important design consideration. Insulation has low conductance and slows down heat gain or loss. Several terms are used to describe various aspects of conduction in a building. The major terms are:

U = conductivity, per inch
U_t = conductivity, total value for unit
R = resistance 1/U per inch
R_t = total resistance for unit

The table below lists R values for commonly used materials.

R Values (per inch)	
concrete	.1
stucco & brick	.2
plywood	1
sawdust	2
fiberglass batts	3
polystyrene	4
exp. polyurethane	6

The resistance of a surface also depends on the air flow across the surface. Air films increase resistance as much as .7 on a vertical surface when it is still. Moving air will drop this to as low as .2. Windows can be a major energy loser because their total resistance may only be 0.9. Even double pane windows at R-1.8 are far below a standard wall at R-13 (with R-11 insulation). Insulated shutters can increase window resistance to as high as R-11 and radically improve the performance of a house with many windows. The conductive heat loss or gain of a building can be calculated by figuring the areas and resistances of all the surfaces and the inside and outside temperature. This is demonstrated in the next chapter.

We can also use conduction to aid in cooling. Conductive cooling most often uses the thermal mass of the earth as a heat sink. One of the most common methods of conductive cooling is the use of an exposed concrete slab thermally coupled to the earth below but insulated on the perimeter. For more conductive cooling the walls can be bermed or the whole house can be buried or bermed.

The potential for conductive cooling can be estimated from the soil temperature at a depth equivalent to that of the exposed surface. The rate of cooling using conductive cooling is low, but combined with solar control and convective cooling it may prove sufficient in many areas. For more detailed calculations of conductive cooling, the soil composition, density, and water content must be known.

Conductive cooling may prove useful almost anywhere in the U.S. where cooling is required. Conductive cooling may also be effectively coupled with a convective loop to provide cool air.

E. Evaporation

One of the simplest methods of natural cooling utilizes the evaporation of water. The areas in the U.S. with the most severe cooling problems are in many cases ideally suited for evaporative cooling.

The latent heat of evaporation is around 60,625 BTU per cubic foot of water; i.e., every cubic foot of water that evaporates provides 60,625 BTU of cooling. As much as 10-15" may evaporate in a month from the typical Class "A" evaporation pan used by the weather service. This translates to a cooling rate of about 2,000 BTU/sq.ft. of evaporating surface per day.

The protection and enhancement of natural wind flow may be very important when evaporative cooling will be used as the windspeed across a surface is a prime factor in evaporation rate. In addition to protecting wind flow, the design of the pond or evaporating surface is critical. If improper edges or lips are included, the wind flow may separate from the surface and markedly reduce the evaporation rate.

At the present time only relatively crude calculations of evaporative cooling potential can be developed for most areas because of limited data on wind speed. If average wind speeds are used to calculate the cooling rate for a "cool pool" house (see Cooling in Chapter 5) the cooling potential will be underestimated because the windspeed is generally highest when the saturation deficit is highest. Conversely, if average windspeeds are used to estimate cooling for a "Skytherm" house it will tend to overestimate evaporation because the pond is exposed when wind speed and saturation deficit are lowest.

If natural evaporation is insufficient, even a small fan can markedly improve cooling performance. If a fan must be very big it may be more advantageous to simply use a manufactured evaporative cooler either to cool the thermal mass directly or to cool the house itself.

3. Heat Storage

The heat storage possible in a house is also important. The ability to store heat is determined by the thermal mass of a building. The massive adobe walls of the old missions are a classic example of the early use of thermal mass. Recent passive solar homes use the thermal mass of water tanks, concrete walls, or earth berms to store "heat" and "coolth" to bring the temperature extremes within the comfort range.

In fact, it is possible to store the summer's heat for use in winter and the winter's cold for use in summer. The first MIT solar house and the more recent "ACES" houses have done exactly that. Studies suggest that this may be economical in the colder areas of the U.S. and Canada. However, in most temperate zone houses storage will be sized to provide only carryover for a few days to reduce the initial cost.

The storage capacity depends in part on several factors including collector system type, storage location, use pattern, local climatic factors, cost, etc. Materials that change phase near room temperature hold promise for increasing storage but few systems using phase change material have been built. The table below presents the various storage properties of commonly used materials.

Properties of Heat Storage Material

Material	Specific Heat BTU/lb./°F	Density lb./ft.³	Heat Capacity No Voids
Water	1.00	62	62
Scrap Iron	0.12	490	59
Scrap Aluminum	0.23	170	39
Concrete	0.23	140	32
Stone	0.21	170	36
Brick	0.21	140	28
Glaubers Salt 80-100°F	-		240
Paraffin	0.7	97	120

There are a number of ways to incorporate added thermal mass in a building or use available mass more effectively. The most obvious, yet often overlooked, mass is that in a concrete slab. All that this requires to be attractive is the use of an integral color, added color, added color and texture, paint, tile, or some other flooring material with good conductivity. A 2" or 3" concrete slab can also be used on the second floor. Tile "seconds" can be very reasonably priced and still attractive.

Added mass can also be added to the walls in numerous ways. On simple walls double sheetrock can be hung easily. With slightly more effort you can plaster walls (up to 1½" in some cases) or add masonry facings. Make sure they are well tied in. A central fireplace (with heatilator and closable doors can also add considerable mass as can a Lorena mud stove.

Water tanks, cans, bottles and drums are also relatively easy to add to a new or existing house. Keep them small enough to move around. Extra bracing may be required on suspended floors or second stories.

The options for adding mass are more varied for a new building. Interior partition walls can be masonry or concrete. Interior tilt up slabs with rock facing might be very attractive and economical. Interlocking stack blocks with surface bonding are relatively easy to work with even for the beginner. Try to spread out the mass because the surface area is important for comfort.

The amount of thermal mass is one of the major determinants of the solar heat fraction. Numerous definitions and explanations of solar heat fraction or percentage solar heating and cooling have been suggested and used. The following is recommended. The percentage of solar heating will include the percentage of heating demand met by both the sun and internal heat gain, as both can be passive sources of energy. The percentage of natural cooling will describe the percentage of the cooling demand not by any of the five methods of cooling. The comfort range will be tempered by climate as experience has shown that the desert heating season begins when indoor temperature drops below 75°F, while in Alaska it may be shirtsleeve weather in the 50's.

Adding thermal mass is easy and well worthwhile but make sure it is safe, particularly in seismic areas. You definitely don't want a 16' x 14" column full of water to fall down. The map included here shows the current estimate of seismic risk in the U.S. The Building Code section on seismic safety is reasonably good but if you are in doubt talk to a structural engineer. It is not difficult to do it right. You might also improve the seismic safety of the rest of the house using the references at the end of this chapter.

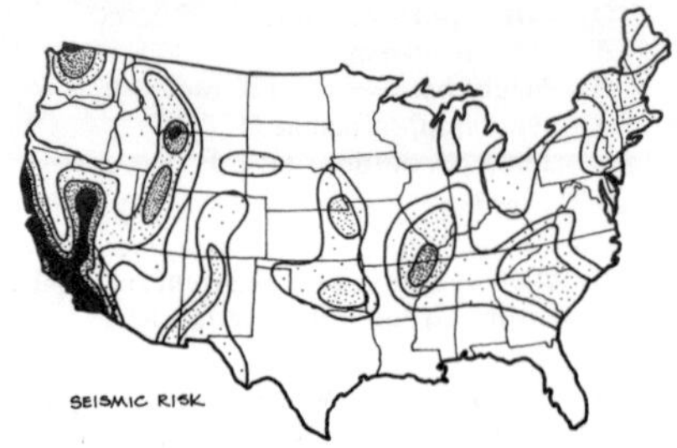

5. Climate and Microclimate

Climate and microclimate are an important part of everyone's life. We are all familiar with the local climate and have a mental image of what our climate is like. These images are often overly general; e.g., hot, dry summers and cold, wet winters and tend to neglect the many small scale microclimatic conditions which vary dramatically. An extensive study of the microclimate of the Neotoma Valley in Ohio vividly illustrates the remarkable changes that may occur in a small area.

Climatic Factor	At 109 Neotoma Valley Stations	Ohio's 88 Stations
Highest annual temperature	75-113°F	91-102°F
Date of Occurrence	Apr. 25-Sept. 19	July 17-19
Lowest January temperature	+14 to -26°F	-6 to -20°F
Frost free period, days	124-276	138-197

Thus, although regional classifications are useful, they may tend to mask the influence of local variations which may significantly increase or decrease individual heating or cooling requirements and design requirements for a passive solar house.

The design and construction of a passive solar system should be based on as accurate a picture of local microclimate as possible. Local measurements may be very desirable for individual houses and are almost mandatory for larger developments. The listing of companies which sell meteorological instruments will help you select a monitoring system for your site.

Much can also be learned from observation and discussion with existing residents. Some of the most critical factors are: topography (sun, cold air drainage, wind, etc.), vegetation (wind, shading, etc.), surface type (water, grass, asphalt, etc.) and structures (shading, wind, etc.).

A description of these factors would be the subject of a book (a rather big book at that) that unfortunately has not yet been written. Only two of the most common concerns are included here.

The first is cold air drainage which can be visualized much like a network of rivers with lakes in every depression. Choose your site wisely to avoid placing your house or garden in a "cold" spot. Buildings and fences can also dam cold air and cause unwanted effects. One chrysanthemum grower in Northern California found this out when he added a long barn across a slope where frost damage had never occurred. Almost all of the plants behind the new barn were frozen while those above the cold air lake line were untouched. It was an expensive mistake.

The second consideration we will touch on here is the variation in wind with site. This can be a concern both for heating and cooling. The nature of local winds will determine the best site. A canyon might be very protected if the dominant winds are cross valley - but if they are down valley, winds may be very strong. Also the desire to have a wind generator may make a ridge top site in a little pass more attractive than a valley site even though high convective heat loss will occur.

Book learning will only help a little bit in this complex field. The best method to learn an area is to spend time out there. Even relatively inexpensive equipment and instruments and plenty of time will enable you to understand your site.

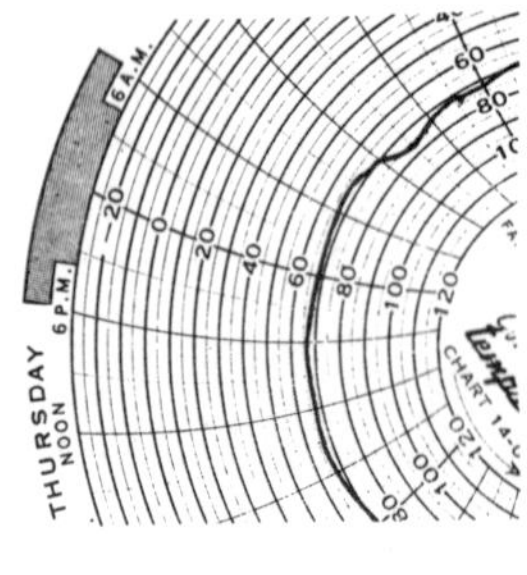

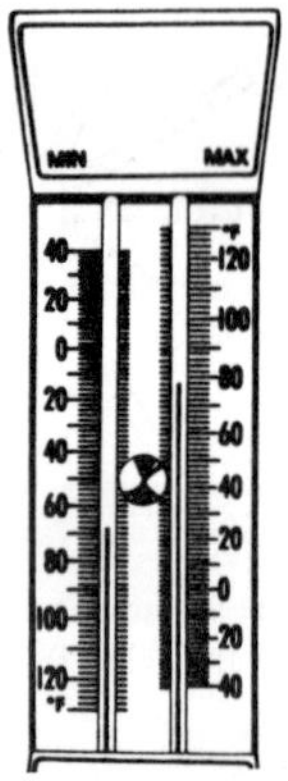

Water Storage Containers	Installed Cost	Source	Notes
Tank	50¢-$1.20	Local welder, or passive tank mfgr.	Aesthetic, easy to install, effective
Drums	20-70¢	drum mfgr. chemical supply, etc.	Cheap, readily available, hard to clean, must be stacked carefully.
Culverts	50¢-$1.20	pipe supply, scrap yards	Strong, attractive to some, where floor space is tight. Make sure installation is seismic safe.
Glass Bottles	free to $2. + gal.	Various	Cheap, readily available, must seal carefully. Installation difficult, seismic problems.
Metal cans	75¢ for 5 gal. cans	Container comp., restaurants (olive oil, etc.)	Good for furniture modules, slow, installation would need bracing for tall stacks.
Plastic bottles and drums	15¢-$1	Various: grocery stores, etc.	Cheap, may provide pleasant light, no rust. Slow to fill, installations weak.

David A. Bainbridge

The following maps are included here to give you a beginning understanding of the climate of the U.S. By all means buy or borrow the books referenced under climate and study them. It will give added richness to your every day life.

MAPS:

Heating Degree Day
Cooling Degree Day
Solar Radiation
Mean Temperature Year

An even more interesting topic is that of climate change. We now know that the climate we have experienced for the last 30 years has been extraordinarily mild. Things are likely to get worse, possibly much worse, and an understanding of climate can help you anticipate how these changes may affect you. Climate changes have brought trouble to many cultures before and we might well question whether we would do any better? Interesting reading and thinking about it. Reid Bryson's book **Climates of Hunger** is particularly well written.

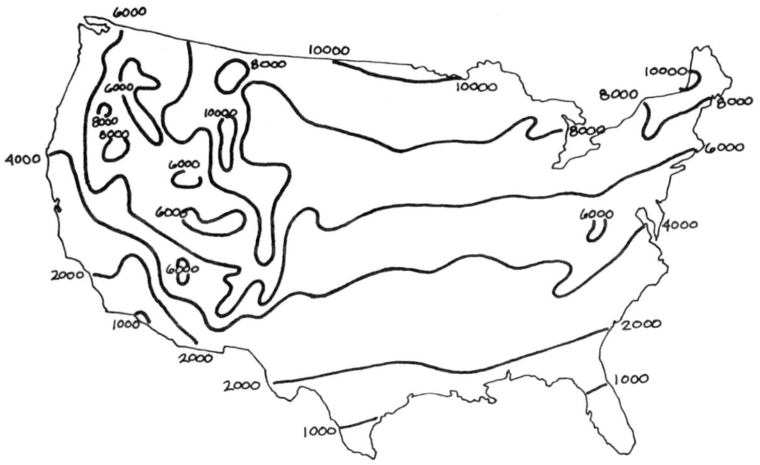

HEATING DEGREE DAYS (65 F)

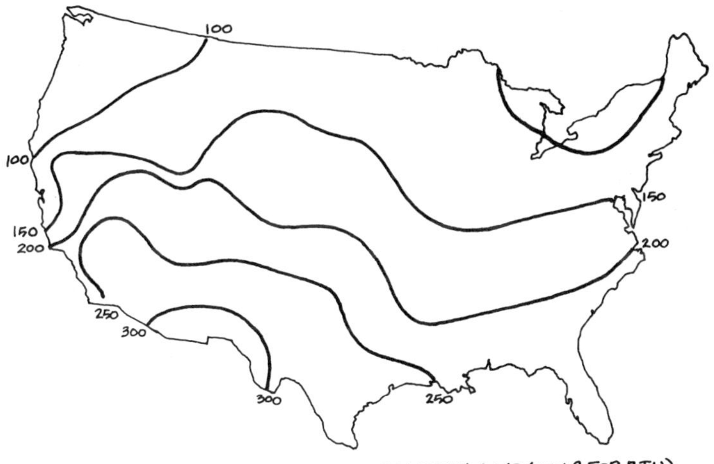

DEC. MEAN DAILY SOLAR RADIATION LYS (x3.69 FOR BTU)

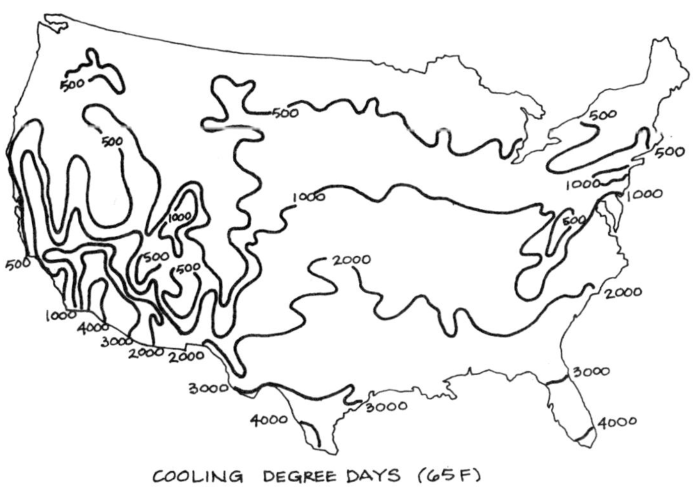

COOLING DEGREE DAYS (65 F)

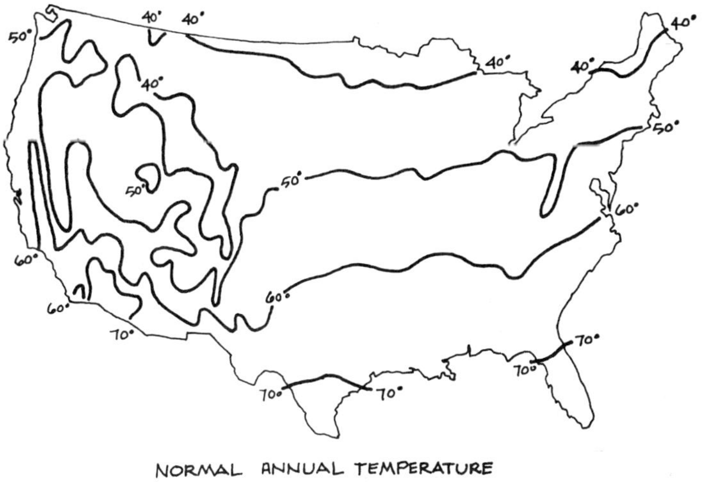

NORMAL ANNUAL TEMPERATURE

5. Human Comfort

The major reason for building a house is to improve the microclimate for man and increase his comfort. Yet all too frequently this result has not been achieved; particularly in houses built before state and local energy conservation codes were enacted. Some houses actually provide fewer hours within the comfort range than would be realized by living outside under a tree. Although these houses may decrease the maximum and minimum temperatures, they may also store heat (or retain the nights cold) long after it has become comfortable outside.

The simple term "human comfort" covers a fairly complex subject, involving all of the methods of heat transfer as well as many psychological and physiological factors too complex to cover in detail here. "Human comfort" is probably best defined as "that range of microclimate conditions under which a person feels good."

This range will vary depending on the type of activity engaged in, age, health, clothing, past experience and adaptation, and body type. For example, a thin, sick, old man may feel cold and uncomfortable in a room described as very comfortable by a young, fit woman. Bedrooms may feel comfortable when much cooler than bathrooms, etc. In fact, human comfort depends on man's thermoregulation system, which is a complex interaction of autonomic and voluntary responses, and mental attitude, and clothes. These govern the rate of heat loss from the body and the rate of heat production as well.

This regulatory system is influenced by many factors including temperature, humidity, radiation, wind, clothing, metabolism, and acclimation. The figure below illustrates the human comfort zone for residents of temperate, cold, and hot zones. As it shows, in hotter climates the comfort zone is displaced up as much as 5° and in colder climates it can be displaced down an equivalent amount. Rather severe tests of human adaptation to temperature have been undertaken and have shown how remarkably adaptable the human body is.

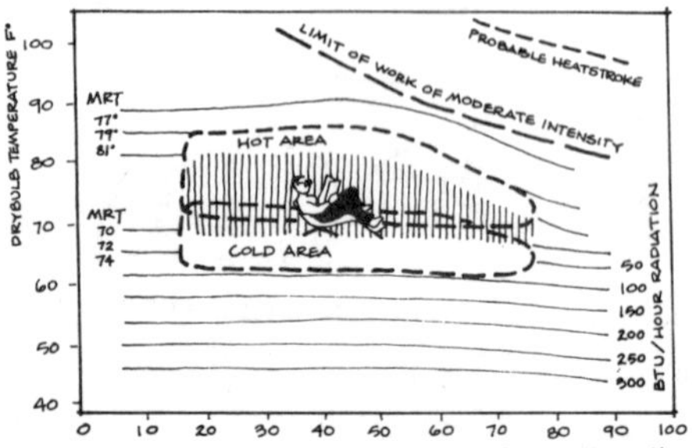

As already mentioned, the comfort range varies with the activities and time of day. When we are sleeping, most people prefer fairly cool temperatures and can be quite comfortable at 55°F (as numerous utilities are now pointing out). This may be very useful in designing passive houses in cold climates where backup systems are needed. If we graph the comfort zone over the day it might look like the diagram below. If the passive system can provide most of the heating, only limited use of the backup heater in the morning may be needed for total comfort.

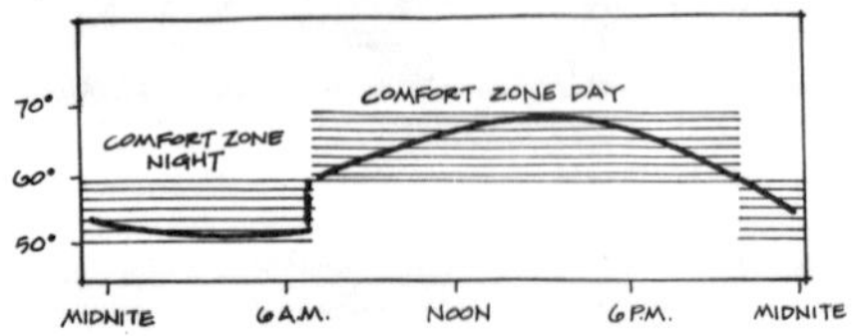

For our purposes, a fairly basic and simplified understanding of the key factors will suffice. As a general rule in dwellings, thermal radiation will be about equal in importance to the combined effects of air temperature, humidity, and air motion. The influence of thermal radiation explains why passive systems are so comfortable.

The ideal system for human comfort is one with radiant temperature in comfort range on all surfaces. This has been known for a very long time and was practiced by the Romans and Chinese in buildings with radiant walls, floors, and ceilings heated by air.

A passive system comes very close to this ideal as well. Although it will typically include only one or two surfaces which are directly heated or cooled, their low and consistent temperatures allow the other surfaces to reach and maintain almost identical surface temperatures. This uniform radiant temperature will provide very comfortable conditions even if the air temperature varies considerably from 72°F. In fact, with proper radiant temperature, it can be quite comfortable with air temperature as low as 60°F or as high as 82°F.

This comfort is readily apparent and is usually one of the first comments made by a visitor to a passive solar house. A look at the more common space conditioning systems will show why this is true.

Forced air systems may maintain the air temperature at the thermostat at 72°F, but they often have trouble reaching all areas of the house with cold pockets and hot pockets as a result. In addition, the hot air quickly rises to the ceiling, leaving the floor too cool and the ceiling too warm with resulting imbalance in radiant temperatures, often 15° or more. And finally, the forced air motion increases convective cooling of occupants as well as providing an irritating, drying, noisy, vibration and breeze.

Radiant heating systems are usually more comfortable, although they also suffer from problems absent in passive systems. The most common variant is the radiant floor, which is usually run with very hot water from a boiler or flat plate, making the floor too hot. This type of system does work fairly well, and is considered one of the more comfortable systems by most people.

Radiant ceiling are often more comfortable than forced air systems, but here the problems are even worse than for a radiant floor. The overheated area is near the head (one of the dominant heat exchange areas for man) and the floor can remain very cool. In addition, air heated on the surface has nowhere to go, so it forms a pool of hot air near the ceiling.

Therefore, it is probably fair to say that passive systems (if well designed) are the most comfortable type of house-heating and cooling systems, and that the people who live in passive houses, and claim they are "More comfortable than ever before" are right.

6. Human Activity

People living in a building affect it's performance in many ways. One of the most important impacts is the heat generated by people, their appliances and their pets. This can either be desirable where heating is needed, or very much unwanted where cooling is the major concern. It will vary from place to place and from one time of year to the next. Design should seek to discover and utilize the best, most cost effective methods of varying internal heating gain to meet the local need.

For a typical family of four the major concerns are the appliances, the refrigerator, people, and the lights. Values determined in a study conducted for D.O.E. are included below:

Weekday Heat Production

People	23,000 BTU/day
Lights	15,000 BTU/day
Appliances (excl.refrigerator)	16,000 BTU/day
Refrigerator	9,600 BTU/day
Total	63,600 BTU/day

This heat may meet most, if not all of the heating in a very energy efficient house in the milder areas of the U.S. even in the middle of the winter. In the summer, however, it may be responsible for much of the cooling demand.

The timing of this heat gain is also worth looking at. The big peak in the evening can provide a valuable pulse of heat to a passive solar house in the winter - after the sun is already down.

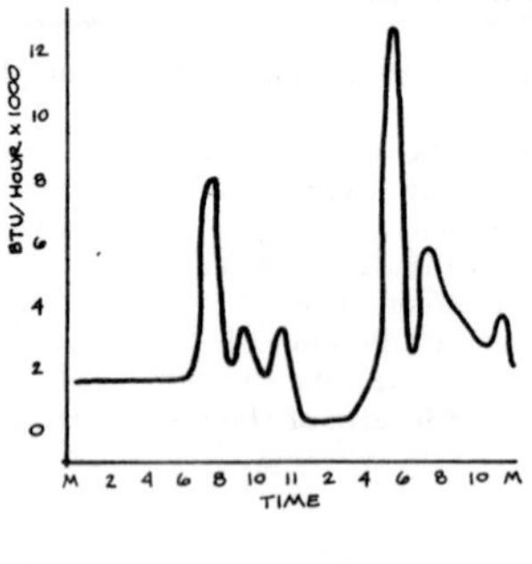

For general purposes a rough estimate of 1.5-2.5 million BTU/month for a family of four will suffice. However, for more detailed calculations of house performance a better estimate is needed. It can be determined by figuring people hours in the house, use of appliances, and other sources of internal gains. The following table lists values of some appliances.

Item	Watts	KWHRS./Month
blanket	190	15
clothes drier, electric heat	4856	86
clothes drier, gas heat	325	
clothes washer,		6
automatic	512	
clothes washer,		9
wringer	275	
coffee maker	800	4
coffee percolator	300-600	12
coffee pot	894	3-10
dehumidifier	300-500	9
dishwasher	1200	50
disposal	445	36
drill, electric ¼"	250	3
electronic oven	3000-7000	5
		100
freezer, food, 5-30 cu.ft.	300-800	30-150
roaster	1320	40
rotisserie	1400	42
sauce pan	300-400	2-10
skillet	1000-1350	5-20
television	300-315	15-30
TV, BW	237	25
TV, color	350	42
record player,		100
transistor	60	
refrigerator		3
water heater	4474	40-400
water pump (shallow)	½ hp	400
water pump (deep)	1/3-1 hp	5-20
		10-60

These use figures can be easily cut 50% with conservation measures.

Another major concern is door opening and closing as it may significantly increase unwanted heat loss or heat gain. This is particularly true for families with many children. A protected air locked door can greatly reduce this problem.

Peoples' desire to sleep in fresh (cold) air can also be a problem. Many people simply can't or won't sleep without having the window open. This "hole" can greatly contribute to heat loss. It may be desirable to insulate and weatherstrip a bedroom from the rest of the house. Or better yet—sleep outside in a porch or covered "bed" room.

David A. Bainbridge

Further Reading (see also Ch. 5)

Radiation

Budyko, M.I. (1958) The Heat Balance of the Earth's Surface, Office of Technical Services, U.S. Department of Commerce.

Geiger, R. (1965) **The Climate Near the Ground**, Harvard University Press.

Sellers, W.E. (1966) **Physical Climatology**, University of Chicago Press, Chicago.

Kondrat'yev, K. Ya. (1965) **Radiative Heat Exchange in the Atmosphere**, Pergamon Press, N.Y.

Meinel, A.B. and M.P. (1976) **Applied Solar Energy**, Addison-Wesley, Menlo Park, CA.

Sun Path

Dufton, A.F. and Beckett, H.E. (1932) "An Instrument for Demonstrating the Apparent Motion of the Sun," **Journal Sci. Instrum**, 9.

Hammond, J. and Long, D. (1976) "The Solar Simulator," **Davis Energy Conservation Building Code Workbook**, Living Systems, Winters.

Parks, R.R. (1950) "The Solar Ranger," **Agricultural Engineering**, March.

Pleijel, G. (1949) **Daylight Investigation**, Report 17, Statens Namnd for Byggnadsforskning, Stockholm.

Staff (1964) "Solar Shadowscape," C.S.I.R.O., Pretoria.

Staff (1974) **Sun Angle Calculator**, Libbey-Owens-Ford Company. ($5.00)

Solar Control

Neubauer, L.W. and Cramer, R.D. (1968) "The Effect of Shape of Building on Interior Air Temperature," **ASAE Transactions** V. 11, No. 4, 537-553.

Neubauer, L.W. (1969) "Southermation: Heat Control for Farm Buildings," **ASAE Technical Reprint PCR**, 69-117.

Neubauer, L.W. (1968) "Effect of Size, Shape, Color and Orientation on Temperature Characteristics of Model Buildings," **Paper No. 68-413, ASAE**, Utah State University, Logan, Utah, June.

Cramer, R.D. and L.W. Neubauer (1958) "Solar Radiation Gains through Directional Glass Exposure," **Heating, Piping and Air Conditioning**, November.

Neubauer, L.W. and Cramer, R.E. (1966) "Solar Radiation Control for Small Exposed Houses," **ASAE Transactions**, V. 9, n. 2.

Anderson, B. and Riordan, M. (1976) **The Solar Home Book**, Chesire Press, Harrisville, N.H.

Hammond, J., Kopper, B. and Bainbridge, D. (1977) "Principles of Passive Solar Systems," in **Davis Energy Conservation Report**, Living Systems, Winters, CA 95694.

Robinette, G.O. (1972) **Plants, People, and Environmental Quality**, N.P.S. Department of Interior, Washington, D.C.

Givoni, B. and Shalon, R. (1962) "Influence of Roof Type and Construction of Indoor Thermal Conditions in Beer Sheva," **Research Paper No. 11**, BRS, Technicon, Haifa.

Hunt, M. and Bainbridge, D. "The Influence of Roof Color and Material on Roof Temperature," in **Indio Energy Conservation Report**, Living Systems, Winter.

Hammond, J., Hunt, M., Neubauer, L.W., and Cramer, R. (1974) **A Strategy for Energy Conservation**, Living Systems, Winters.

Neubauer, L. and Walker, Harry B. (1961) **Farm Building Design**, Prentice-Hall Inc., Englewoods Cliffs, New Jersey.

Olgyay, Victor and Aladar (1976) **Solar Control and Shading Devices**, Princeton Univ., Princeton, N.J. $7.50.

Thermal Radiation

Sellers, W.D. (1966) **Physical Climatology**, University of Chicago Press, Chicago.

Budyko, M.I. (1958) **The Heat Balance of the Earth's Surface**, Office of Technical Services, U.S. Department of Commerce.

Brooks, F.A. (1959) **An Introduction to Physical Micrometeorology**, University of California, Davis.

Pleijel, G. (1954) **The Computation of Natural Radiation in Architecture and Town Planning**, Statens Namnd for Byggerandstorsking, Stockholm.

Neubauer, L. and Cramer, R. (1965) "Diurnal Radiant Exchange with the Sky Dome," **Solar Energy**, v. IX, n. 2.

Kondrat'yev, K. Ya. (1965) **Radiative Heat Exchange in the Atmosphere**, Pergamon, N.Y.

Reitan, C.H. (1959) "Distribution of Precipitable Water Vapor over the Continental U.S." **Bulletin American Meteorological Society**, v. 41, n. 2, pp. 79-98.

Geiger, R. (1965) **The Climate Near the Ground**, Harvard Univ. Press.

Goss, J.R. (1956) "Cooling of Water by Nocturnal Radiation and Evaporation," M.S. Thesis, Agricultural Engineering, UC Davis.

Hay, H. and Yellot, J.I. (1959) "International Aspects of Air Conditioning with Movable Insulation," **Solar Energy**, v. 12, pp. 427-438.

Clark, G., Loxsom, F. and Niles, P. (1977) "Roof Pond Cooling vs. Active Solar Cooling in a Subtropical Marine Climate," presented at the Helioscience Institute, Palm Springs.

Barker, Sir Robert F.R. S. (1775) "The Process of Making Ice in the East Indies," **Proceedings of the Royal Society**, Redde, March 9.

Niles, P., Haggard, K. and Hay, H. (1976) "Nocturnal Cooling and Solar Heating with Water Ponds and Movable Insulation," **ASHRAE Transactions**, v. 82, part 1.

Pittenger, A.L., White, W.R. and Yellott, J.I. (1978) **The Energy Roof**, Solar Building Systems, Tempe, AZ.

Bainbridge, D. (1976) "The Indio Cool Pool Experiment," **Alternative Sources of Energy**, June No. 32.

Plumb, J. (1978) "Quarterly Report: ERDA Cool Pool Investigation," Living Systems, Winters.

Bainbridge, D. (1978) "Natural Cooling in California," **Proceedings ISES** Conference, Denver, Colo.

Hay, H. and Yellott, J.I. (1969) "Natural Air Conditioning with Roof Ponds and Movable Insulation," **ASHRAE Transactions**, pt. 1, v.pp. 75-177.

Hammond, J. (1976) "The Winters House," **Passive Solar Heating and Cooling: Conference and Workshop Proceedings**, Los Alamos, NTIS-LA 6637-C.

Givoni, B., Paciuk, M. and Weiser, S. (1976) **Natural Energies for Heating and Cooling of Buildings, Analytical Survey**, BRS. Technicon, Haifa.

Kelly, C.F., Bond, T.E., and Ittner, N.R. (1957) "Cold Spots in the Sky May Help Cool Livestock," **Agricultural Engineering**, October.

Abraham, F.F. (1960) "Determination of Long-Wave Atmospheric Radiation," **Journal of Meteorology** v. 17, pp. 219-292.

Convective Cooling

ASHRAE (1977) **Handbook of Fundamentals**, ASHRAE, N.Y.

Pearson, J. and Terry, C. (1978) **Hawaii Home Energy Book**, University Press of Hawaii.

Strock, C. and Koral, R. (1963) **Handbook of Air Conditioning, Heating, and Ventilation**, Industrial Press, N.Y.

Hammond, J., Hunt, M., Bainbridge, D., *et al.* (1976) **Site 1: State Office Building**, Living Systems for the Office of the State Architect.

Hunt, M., Bainbridge, D. *et al.* (1978) **A Simple Solar House for California**, Davis Alternative Technology Associates/Solar Office, California Energy Commission, Sacramento.

Olgyay, V. (1963) **Design with Climate**, Princeton University Press.

White, R.F. (1945) **Effects of Landscape Development in the Natural Ventilation of Buildings and their Adjacent Areas, Research Report No. 45**, Texas Engineering Experiment Station, Reprint (1976).

Baer, S. (1975) **Sunspots**, Zomeworks. Albuquerque, N.M.

Conductive Cooling

Wells, M. (1977) **Underground Designs**, M.W., Box 1149, Brewster, MA 02631.

Givoni, B. (1976) "Modifying the Ambient Temperatures of Underground Buildings," Presented at the D.O.E. **Conference on Underground Buildings**, Fort Worth, Texas, May.

Meter, K. (1977) "An Interview with Tom Bligh," **Alternative Sources of Energy**, No. 26 June.

Moreland, F.L. ed. (1976) **Earth Covered Buildings for Energy Conservation**, Conference Proceedings, University of Texas, Arlington.

Underground Space Center (1978) **Earth Sheltered Housing Design**, U.S.C. Univ. of Minnesota, Minneapolis.

Evaporative Cooling

Bahadori, M.N. (1978) "Passive Cooling System in Iranian Architecture," **Scientific American**, Feb.

Watt, J.R. (1963) **Evaporative Air Conditioning**, Industrial Press, N.Y.

Thermal Storage

National Concrete Masonry Institute. (1977) Several Pamphlets on Passive Solar Design Using Concrete, 6845 Elm St., Suite 711, McLean, VA 22101 including: "New Insights into Energy Use and Conservation in Structures." 40¢.

Balcomb, Hedstrom, and McFarland (1977) "Thermal Storage Walls for Passive Solar Heating Evaluated," **Solar Age**, August.

Bainbridge, D. (1978) "A Passive Solar Home for the Mass Market," **Alternative Source of Energy**, August.

Trombe, F. et al. (1977) "Concrete Walls to Collect and Hold Heat"...**Solar Age**, 2(8): 13-19.

Anderson, B. (1977) "Designing and Building a Trombe Wall"...**Solar Age**, 2(8): 25-28.

Wright, D. (1976) "Adobe: Designing for Passive Solar Heat in the Southwest," **Solar Age**, 1(8): 10-15, July.

Ohlson, Keith (1978) **Passive Solar Energy Applications of High Mass Walls in Architecture**. Senior Project California Polytechnic State Univ.

Berkeley Solar Group (1978) **Passive Solar Building with Concrete Masonry**, Concrete Masonry Association of California at Nevada, Sacramento, CA.

Kern, Ken and Magers, Steve (1976) **Stone Masonry**, Owner-Builder Publication.

Army (1975) **Concrete, Masonry, and Brickwork**. Dover.

Staff (1977) **Masonry**, Time-Life.

Bolt, Bruce (1978) **Earthquakes: A Primer**, W.H. Freeman, $12.00.

Yanev, Peter (1974) **Peace of Mind in Earthquake Country**, Chronicle Books, San Francisco, $4.95.

Climate

Environmental Data Service (1975) **Climatic Atlas of the U.S.**, U.S.G.P.O.

Sellers, William (1965) **Physical Climatology**, University of Chicago Press, Chicago.

Geiger, Rudolf (1965) **The Climate Near the Ground**, Harvard University Press.

Rosenburg, Norman (1974) **Microclimate: The Biological Environment**, John Wiley and Sons.

Sutton, O.G. (1953) **Micrometeorology**, McGraw Hill, N.Y.

Staff (annual) **Local Climatological Data: Annual Summaries**...U.S. Deparment of Commerce, National Oceanic and Atmospheric Administration, National Climatic Center, Federal Building, Asheville N.C. 28801.

Comfort

Baruch Giovoni (1969) **Man, Climate and Architecture**, Elsevier.

Egan, David (1972) **Concepts of Thermal Comfort**, Tulane University of Architecture.

Fanger, P.O. (1970) **Thermal Comfort**, McGraw-Hill, N.Y.

ASHRAE (1977) **Handbook of Fundamentals**, ASHRAE, N.Y.

Winslow, C.E.A. and Herrington, L.P. (1949) **Temperature and Human Life**, Princeton Univ. Press.

Fitch, J.M. (1972) **American Building: The Environmental Forces That Shape It**, Houghton-Mifflin.

Human Activity

ASHRAE (1977) **Handbook of Fundamentals**, ASHRAE, N.Y.

BHK & R Associates (1978) **Minimum Energy Dwelling Workbook**, DOE, San/1198-1.

Chapter 5: An Introduction to Passive Solar Heating and Cooling Design

"Before you put on your overcoat you button up your sweater."

—Marshall Hunt

This chapter describes how passive solar heating and cooling systems work. It includes methods that will help you size a system to fit local climatic conditions. The first part of the chapter describes how to upgrade the shell and tune up the various systems of the house to make it more efficient—before more sophisticated heating or cooling systems are added. Although many of these energy conservation techniques and strategies can be applied most easily in new construction, many can also be included in remodels or retrofits.

1. Energy Efficiency

Improving the energy efficiency of a house beyond the standard level is remarkably easy. In most areas it is reasonable to expect at least a doubling in the energy efficiency of the house—with only minor increases in cost. The following steps should be included: A) orientation, B) insulation, C) weatherizing, D) shading, E) modifying the environment, F) appliances, and G) lighting.

A. Orientation

Good orientation is essential for an energy efficient house. The effects of bad orientation can be remedied to some extent—but doing so can be expensive. The best orientation is for the house to be longer East-West than it is North-South, with a major yard to the south, and most of the windows facing south and a few facing north, east or west. This orientation will maximize solar heating in winter and minimize summer overheating by making best use of the difference in sun path. Even a house with less than ideal shape will work well if the glazing is predominately on the south side.

The impact proper glazing orientation can have even in a mild climate is shown in the following comparison of a typical 3-bedroom house in Sacramento with three sliding glass patio doors on the major wall and a three foot overhang.

	Major Wall and Windows Facing	Total BTU/gain	Result
June 21	West	120,000	1/3 as much cooling required
	South	36,000	
December 21	West	64,000	4 times as much solar heating
	South	230,000	

B. Insulation

The second step in building or remodelling for energy efficiency is reducing unwanted conductive heat loss (or gain). Insulation is the key—not only for the walls and ceiling, but also for the foundation or slab perimeter, the windows, and the people inside.

a. The Walls and Ceiling

Insulating the walls and ceiling has become standard practice. However, this is a recent occurance and much of the existing housing stock in milder areas still has no insulation. As a result the typical house in Los Angeles (2000 heating degree days) has been using as much energy for heating as a house in Michigan (8000 heating degree days).

Although many articles have recently been published which recommend certain minimum levels of insulation as being most economical, these articles have all made several questionable and in the case of passive solar houses, incorrect assumptions. First, they assume relatively high infiltration which will not occur with passive design; second, they assume that windows are not fitted with insulated shutters or drapes; and finally, they assume relatively stable energy prices and/or continued assured delivery of fuels. A general rule of thumb to folllow when adding insulation is to install as much as you can.

The values below are desirable:

Highest value of either Heating or Cooling Degree Days

		Wall	Ceiling
Under	2,000	4-11	R-19
	4,000	R-19	R-30
	8,000	R-19	R-40
Over	8,000	R-24	R-50

Recommended Insulation

The only exception to this rule is for south walls of buildings built with adobe, stone, or concrete in temperate areas with good winter sun. These may only have a static R-value of 3 or 4. However, due to their thermal storage these walls have a much higher effective "R" value and in fact act like a window and will show a net heat gain for the winter.

The types of insulation most commonly used include: blown in fiberglass (R=2.25 inch), fiberglass batts (R=3 inch), blown in rock wool (R=2.7 inch), blown in cellulose (R=3.7 inch) and expanded plastic foam (R=4-5). The most attractive in terms of environmental impact and net energy cost are the cellulose and rock wool insulations. Fiberglass may take many years to pay for the energy used in manufacturing, rock wool considerably less time, and blown in cellulose probably only a few months. Make sure that your cellulose is properly fireproofed. One of the advantages of cellulose insulation is that you can make it yourself. **Mother Earth News** presents the details in the Nov./Dec. 1977 issue, "How to Make and Intall Your Own Insulation for 5¢ or less a square foot."

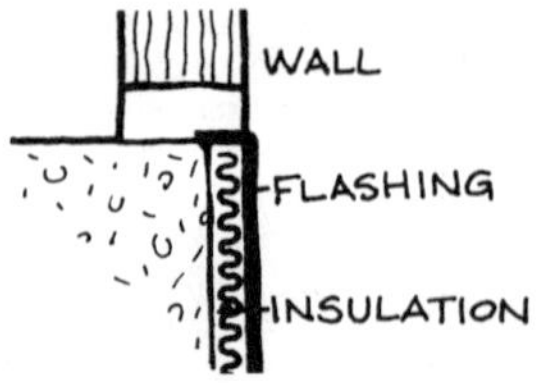

b. Floors and Foundations

The insulation for perimeter slabs, foundations, or floors should also be very good. The slab or foundation of a house can pump warmth out in the winter and in during the summer. The insulation should extend far enough into the ground to act as a good barrier to heat flow. In warmer areas 1-2' will suffice. In very cold areas it should go much deeper, at least to the frost line.

c. Doors

Doors should also be insulated. An uninsulated door may have an R-value of only 1 or 2 unless an insulated door is specified or built. Insulated doors can be purchased or built using the techniques described in this book for making insulated shutters for windows.

d. Windows

Windows, however, are the single most important area where insulation is needed. Single pane glass has an R-value of only 0.9. Double pane is better at about R-1.8. On a well insulated house uncontrolled double pane windows may account for 30-40% of the heat loss. Clearly, some form of glazing control will be effective in reducing unwanted heat loss and heat gain. Either insulated shutters or drapes will fill the bill.

The choice for each area will depend on aesthetic and functional requirements. Rigid shutters can be particularly effective. They can easily have R-values of 10 and even R-15 is not terribly difficult to achieve. The accompanying diagram shows how a rigid shutter can be built. One of the advantages of shutters is that they can be weatherstripped to seal tightly and reduce infiltration (the next section).

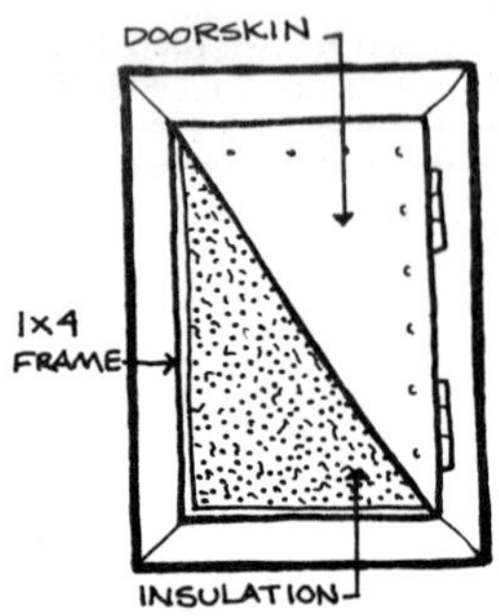

Several tricks can be used to make insulated shutters more desirable and attractive. One of the common concerns is that they block out all of the view and light. A simple small double pane window can be installed in the shutter to overcome this objection. Or alternatively translucent materials can be used. A second concern involves summer ventilation. Privacy is still desired, yet ventilation is needed. The solution is simple—make up a louvred shutter to fit the same space. In spring pull the hinge pins on the insulated shutters and hang them in the closet or garage and replace them with the louvred shutters.

Insulated drapes or curtains are another possible situation. These can be made or purchased at costs comparable to traditional drapes. To work well they should be sealed at the edges, top, and bottom. Several methods can be used to accomplish this seal, including tracks, magnetic strips, weighted bottoms, and valenced tops.

Insulation can be included with synthetic insulation such as Polarguard™, Fiberfill II™, or Thinsulate™ or by using several layers of fabric. Several very attractive drapes are now manufactured using these different approaches. Or you can do-it-yourself following the plan included here.

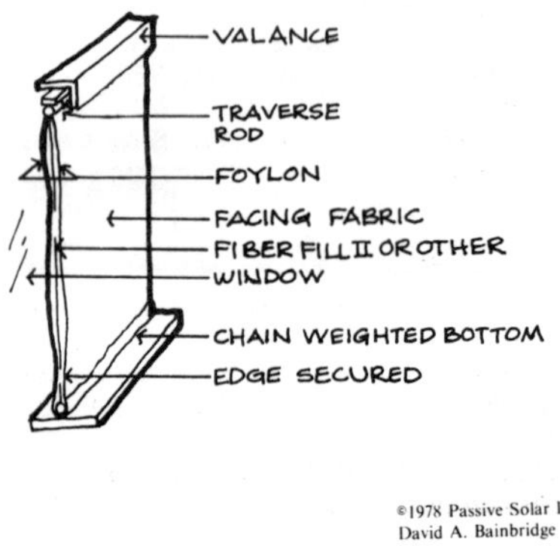

The simplest solution is a cross between a shutter and curtain and can be made quickly with simple tools at a cost of about 35¢ per square foot. It uses a cheap sleeping bag as the cover for a simple wood frame. Make the wood frame about ½" short and then cut the sleeping bag to fit around it and staple it on the back, or better yet, fasten it down with strips of wood. The sleeping bag material will act as a gasket to hold it in and prevent infiltration.

e. Headers and Wall Members

Although traditionally headers (the beams over windows and doors) and wall members are not insulated it makes sense to insulate the wood pieces in a wall. A 4" x 8" or 4" x 10" header over a door or window will have an R value of only 4. These uninsulated wood pieces act as small windows for heat loss. The framing contributes more heat loss if the studs are on 16" centers as over 15% of the wall section may be wood. The use of 2" x 6" walls on 24 inch centers can reduce the wood area to about 10%. Use of 2" x 6" studs also allows 2" of foam insulation board to be placed on the outside of the 4" wide headers to bring the total R-value to 14.

The insulation values of the various parts of the house can be added up to determine the overall conductance of the house. To figure this out calculate the area of all the different parts of the house exposed to the outside. Divide each area by the R-value of the insulation and other materials in the wall and then add up the totals for an idea of the house's efficiency.

The value of careful insulation can be seen most clearly with an example using a typical three bedroom house with a floor area of 1600 square feet, window area of 250 square feet, door area of 42 square feet, net wall area of 1200 square feet and a slab edge of 160 feet. The conductance of this house in four variations is shown below: As a passive solar house, a conventional house built with typical practice, and as a similar house built in 1965 and in 1940.

	1940	**1965**	**Typical**	**Passive**
Roof conductance 1600 sq. ft	R-3=500	R-11=145	R-19=84	R-30=53
Wall conductance 1200 sq. ft.	R-3=400	R-7=171	R-11=109	R-19=63
Window conductance 250 sq. ft.	R-1=250	R-1=250	R-2=125	R-11=23
Door conductance 42 sq. ft.	R-2=21	R-2=21	R-2=21	R-7=6
TOTAL CONDUCTANCE	1171 BTU/hr.	587 BTU/hr.	339 BTU/hr.	145 BTU/hr.

To really appreciate the importance of this improvement, assume that the design temperature is 10°F and the desired indoor temperature is 70°F. The total heat loss per hour then is shown below, including slab edge losses calculated using the ASHRAE procedures:
Passive House: (145 x 60) = 8700 BTU + (160 ft. x 10 BTU/ft.) = 10,300 BTU/hr. loss
Typical House: (339 x 60) = 20340 BTU + (160 ft. x 50 BTU/ft.) = 28,340 BTU/hr. loss
1965 House: (587 x 60) = 35220 BTU + (160 x 50 BTU/ft.) = 43,220 BTU/hr. loss
1940 House: (1171 x 60) = 70260 BTU + (160 x 50 BTU/ft.) = 78,260 BTU/hr. loss

As you can see conducted heat loss is less than ¼ of that in a typical house built only 10 years ago. This is not the whole story though, because infiltration is also an important factor in the thermal performance of a house.

f. People

With particularly difficult retrofits or rentals it may be best to spend more money on insulating the people rather than the house. The increasing popularity of backpacking and mountaineering has led to the development of many extremely comfortable, lightweight, and effective products for staying warm while working and sleeping. Instead of spending $1500 to insulate the walls on an older house you might be better off to spend $400 to equip the family with good thermal underwear, hats, booties, clothes, and comforters. Only someone who has slept under a down comforter can fully appreciate the comfort possible even if the temperature drops into the 40's (30's or 20's).

The value of warm clothes can be seen most clearly if we again look at human comfort. The major variable here is the clo or the clothing/skin value figured for the body.

The clo values for typical clothing ensembles are listed below:

Clothes	Resistance in m²/hr/°C/kcal
Nude	0
Tropical clothes - shorts/sandals/short sleeve	0.3
Light summer clothes - light pants/short sleeve	0.5
Typical business suit	1.0
Heavy business suit	1.5
Wool pants, shirt, and coat	2.0
Pile ensemble	3-4+
Down ensemble	3-4+

We can see that we can reduce personal heat loss by ½ to ¾ simply by improving the insulating value of our clothes.

C. Weatherizing

Infiltration losses are as important as conductive losses and careful weatherizing is necessary. This includes both the obvious problems of weather-stripping doors and windows and also the more general problems of caulking and sealing of building joints, access holes, and any other areas where unwanted infiltration occurs. Infiltration may account for almost ½ of the heat loss in a well insulated but poorly weatherized house.

The infiltration rate on a typical house is around 1.5 air changes/hour. It can go up significantly if a high wind is blowing. However, with very careful attention to detail this can be reduced to about 0.5 air changes/hour. The Swedish have reduced infiltration in some of their houses to 0.2 air changes/hour.

The importance of this reduction can be seen by looking at the 1600 sq. ft. house again. Assuming the same design temperature the heat loss per hour will be:

Type House	**(Airchange x Volume)**	**(Heat Capacity of Air)**	**(Temp inside - Temp outside)**	**= BTU/heat loss hr**
Typical house:	(1.5 x 12,800 cu. ft.)	(.018)	(70-10)	=20,736 BTU/hr
Passive house:	(0.5 x 12,800 cu. ft.)	(.018)	(70-10	= 6,912 BTU/hr
Swedish house:	(0.2 x 12,800 cu. ft.)	(.018)	(70-10)	= 2,764 BTU/hr

Compare this to the conducted heat loss and it will become apparent why insulation alone is not enough.

For a new house thorough and efficient weatherizing is possible. It begins with the junction between the plate, or bottom of the wall and the foundation. Here a layer of mastic or sealing tape should be placed before the plate is bolted down. Second, as framing continues stuff cracks in corners and joints with insulation. Third, as the windows go in insulate and caulk around the window-rough frame intersection. The tightest windows available should be used. Insulated shutters can reduce infiltration considerably if they seal tightly.

After the insulation is installed go around the house and make sure all the gaps are filled in. The areas where insulation installers are most lax are often the areas where the worst infiltration will occur. Service entry for gas, water, and particularly wiring are all critical. After the wiring and plumbing are done seal the holes where the wires and pipes enter the walls from the attic. These often serve as an entry for cold air. Also, as the electric boxes are installed seal around the wire entry into the box (with a non-flammable insulation). Try this test on a windy day to see why. Blow smoke toward an electrical outlet and see how it is blown away. In most cases you will find a jet of cold air coming in.

After the sheetrock is up you should go around the house and seal any gaps in the sheetrock. A bead of caulk along the bottom of the walls will help. So will sealing around heater boxes, fixtures, etc. Every little bit helps. Smoke can also be used to test the tightness of the seal around the baseboard.

As the doors are hung get interlocking weatherstrip installed. This may cost $60 or more per door but it is durable and effective. Or use one or several of the alternatives such as spring strips, foam, or one of the other types of weatherstripping.

The biggest problem with doors, however, is not when they are closed - but when they are open. An airlock or vestibule is the only way to really control this problem. The garage in many cases can be used as the airlock. Or a "mudroom" or cold room can be built into the main entryway. The winter entry should also be sheltered from the worst winter winds.

The final area of concern is the various flues and stacks that penetrate the roof. These may total as much as 60 square inches of hole in the roof, the area where heat loss is most rapid. No solution is applicable to all of these but take a good look at the system and figure out where you can install better dampers and weatherstripped lids and doors.

As the house is made tighter and tighter the control of indoor air contaminants becomes more important. The use of pilotless stoves and water heaters and well ducted furnaces and wood stoves using cold air intakes can minimize this problem. In addition, a ventilator with heat exchanger may be used to prewarm fresh air, with dirty air as it is exhausted. These can be made or purchased, see Chapter 11.

Weatherizing an existing house is more difficult but significant improvements can usually be made. Windows and doors are often the worst culprits in an older house. Careful caulking and weatherstripping are the first steps. Storm windows, also carefully sealed, will be very helpful and also help reduce conductive losses. Tightly sealed insulating shutters and drapes will further reduce infiltration around the windows.

The doors should be weatherstripped with interlocking weatherstripping if you can afford it. If you can't, then diligent use of foam insulation and a plastic base strip may achieve reasonably performance. It may also be possible to make a hallway or back room an airlock entry.

The walls on an older house may be very leaky if there is no insulation in the walls or caulk under the plate. Insulation will help if you can afford it. Caulking seams and joints may also help. Blow some smoke near the baseboard to see how much air is blowing under the plate. If it is bad (and it may surprise you) take off the baseboard trim and apply a bead of caulk along the base of the wall.

Get up in the attic and seal cracks and holes where cold winter air may be entering the house. Use discretion when insulating near light fixtures or wiring. Be particularly careful if the insulation/seal might cause overheating or fire problems.

Many old houses unfortunately have fireplaces. These can be a very effective thermal chimney for pumping nice expensive hot air out into the cold. A damper is a poor solution at best. See the section on heating backup for more information on retrofits for fireplaces.

These relatively inexpensive steps should help make your house much tighter and less expensive to heat or cool.

D. Shading

The importance of effective shading is best illustrated with an example. In the Central Valley near Sacramento 55 square feet of west facing glass will add 55,000 BTU to a house during the afternoon. A ton of air conditioning is required to offset this gain. And to run the airconditioner at the peak period of electrical use almost 2 kw of generating capacity are required.

Shading or solar control is relatively easy if the house is oriented properly. Overhangs and wing walls can shade most south windows. Those windows that cannot be shaded in this manner may be shaded with any of the exterior shade screens. The problems and potentials of Solar Control are described in the section on Passive Cooling.

E. Modifying The Environment

Very substantial savings in energy use for heating or cooling may be realized by altering the environment the house sits in. Three possibilities are particularly attractive: landscaping, berming or undergrounding, and building a house within a house.

a. Landscaping

Landscaping and site design can modify the local microclimate to a surprising degree. In summer, good shade trees can reduce neighborhood temperatures 10°F. This may result in an energy saving of up to 30% or more. Landscaping can also be used to channel cooling breezes into a house in the summer.

In winter a good wind break may reduce energy use 20% or more. To be effective it should be relatively dense and tall. It works by reducing both infiltration and conductive losses. An example is shown in the accompanying drawing.

The placement of buildings and fences can also be used to improve the microclimate. They can be set to minimize winter winds and funnel cooling summer breezes in many areas.

b. Berming or Underground

Earth berms or undergrounding can reduce energy use for space conditioning as much as 50%. It is most attractive for areas with either severe cooling or heating problems but will probably be used increasingly even in relatively moderate climates.

Even 7-8" of dirt in a berm or sod roof will act as a good thermal barrier by reducing daily extremes and keeping the temperature near the daily average. More soil will move the averages toward the weekly, monthly, or yearly average.

Undergrounding should include careful insulation near the exposed surfaces or heat wicking may occur. The well insulated bermed or underground house is efficient because the interior space is now seeing an exterior temperature of 60°F instead of 10°F and infiltration is greatly reduced.

Heat loads can be calculated using the techniques described in **Earth Sheltered Housing Design**.

photo: Karen Terry

c. House Within a House

The final environmental modification discussed here has been developed and promoted by Lee Porter Butler and deserves wider attention. It essentially builds a weatherproof shell around the house with convective flow around the interior house. Thus, the heated or cooled space may see a temperature difference of only 10°F instead of 50°F. It also serves to reduce infiltration. Energy savings of 50% may be realized this way. For more details see the Smith House Solar Greenhouse example.

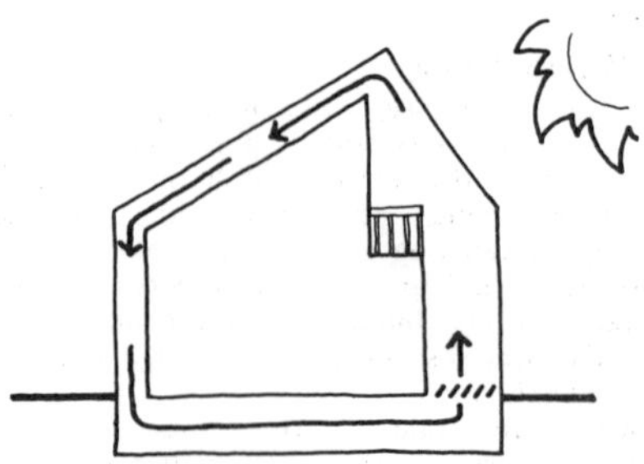

A similar effect can be realized in other houses by carefully zoning the house. An exposed bedroom or porch may be left unheated in winter to reduce heat loss. Or bedrooms can be left largely unheated to provide sleeping with the windows open without expensive heat loss.

At the most refined level the conditioned space may be reduced to a small well insulated room or canopied bed. This is particularly attractive for renters in very inefficient buildings or building where the heating or cooling equipment works poorly, or not at all. Not surprisingly a 4' x 6' space 4' high can be heated or cooled with very little energy, using very inexpensive equipment.

F. Appliances

The energy efficiency of appliances can also play a large role in the energy use for heating and cooling. The choice and use of hot water heater, refrigerator, stove, heating and cooling system can all be very important. All should be chosen and used wisely.

The gas or electric hot water heater should of course be used only when solar water heating with a passive or active solar water heater is insufficient. The tank should be extremely well insulated - using a retrofit jacket. The temperature should be kept as low as possible. All too often they are set at 160°F instead of 120°F. This reduction in temperature makes it less likely someone will be scalded and it may also reduce energy use 20-30%. A setback thermostat is available for electric waterheaters. It will pay for itself very quickly. Pipe runs should be kept as short as possible and well insulated.

An instant hot water heater may be very effective in new or retrofit situations. A 1 quart boiling water tap like the In-Sink-Erator Hot Tap can reduce inefficient heating of water on the stove. It can be switched and turned off when not in use. Additional savings can be realized by using hot water wisely. For example, rather than heating water for each cup of tea as you drink it - make a potful and store it in a Thermos bottle. If you take a hot shower put in the plug so the hot water sill sit in the room and warm it up - rather than warming up the sewer. Think about how you can use your hot water most effectively!

The refrigerator is also important. In an efficient passive solar house with solar hot water, the refrigerator may use more than ½ of the energy used in the house. Recent models are particularly inefficient and may use more than 200+ kwh month to self defrost, keep the butter soft with its own heater, and prevent condensation on the outside with electric heaters because the insulation is too thin.

Energy use of 40-50 kwh month is more reasonable and can be achieved using an older model (around 1950) or with several refrigerators now on the market. (Amana, White Westinghouse, and others). Use of less than 20 kwh month should be attainable with either a horizontal refrigerator built to todays standards or a well designed built in refrigerator.

The heating and cooling system should also be chosen and maintained for maximum efficiency. These concerns are addressed more fully under the Backup section which follows the sections on Passive Heating and Cooling. It is worth noting that European equipment is in many cases twice as efficient as that commonly available in the U.S.

G. Lighting

Careful task lighting, efficient lights, and day lighting can greatly improve the energy efficiency of a home lighting system as well as a commercial building. Fluorescent lights are well suited for areas where exposure will be limited. More information on lighting is included in Chapter 8 on Daylighting.

H. Summary

By taking these simple steps toward energy efficiency house energy can easily be reduced to ½ that used in the typical house today. This enables the passive heating or cooling system to meet more of the remaining demand efficiently and economically. Remember these steps: orientation, insulation, weatherization, shading, environmental modification, energy efficient appliances, daylighting. Then consider passive solar heating and cooling.

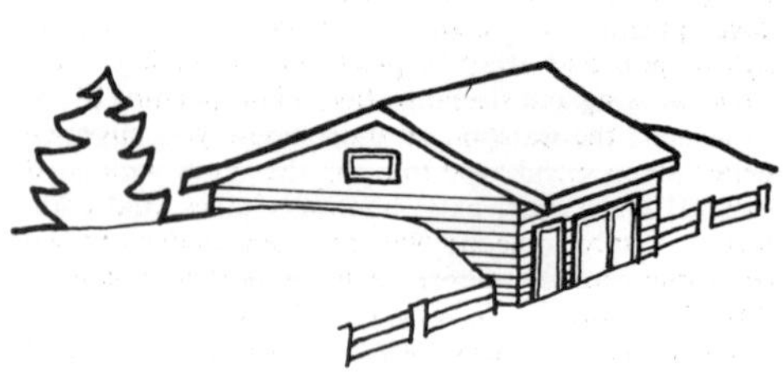

2. Passive Heating

Passive solar systems are very effective for heating. They differ widely in configuration, performance, and cost, but do have several features in common. These include: glazing to admit the sun's energy, glazing control (insulated drapes or shutters) to reduce unwanted heat loss or heat gain through glazing, and thermal mass to store the sun's energy. The following rules apply to all passive solar heating systems.

SIX RULES FOR LOW COST PASSIVE HEATING

i. Make sure the house shell is weather tight and well insulated.

ii. Make sure the house is oriented properly, with most windows on the south.

iii. Use the minimum amount of glazing necessary for heating.

iv. Make glazing control simple, reliable, effective and convenient to use.

v. Use the simplest, smallest, and most economical method of providing needed thermal mass.

vi. Consider cooling demand as well and integrate design to achieve lowest cost heating and cooling.

With these rules in mind, let us look at the different types of passive heating systems. Many needlessly complex classifications for passive solar heating systems have evolved. They have been over-developed to the point where they almost provide a category for every house in the U.S. Unfortunately they have ignored much of the international work that would require new classifications. A simplified description is used here for the following passive house types:

A) The Simple Passive House - A suntempered standard frame house

B) Free standing mass element

C) Integral Massive Element - wall or roof

D) Underground or Massive Construction

For each of these classes there are almost unlimited variations in design, performance and cost. Some of the more common and promising types are described here. After a description of passive heating and cooling the systems are introduced as they work for both heating and cooling. In addition, the detailed examples in Chapter 6 describe specific house designs in greater detail. Before we get to that, however, the basics of passive heating will be described, along with several rules of thumb and simple methods to calculate required window area and mass. The performance of these systems can also be augmented using reflectors, and in some cases, fans, as described in Chapter 4.

A. The Simple Passive House

The standard frame house can function fairly effectively as a passively heated solar house if it faces south and has good glazing control. This type of house is often referred to as a sun tempered house. With good solar exposure and careful operation of thermal shutters and drapes, a lightweight frame house may meet from 10 to 50% of its heating demand using the sun. With an exposed tile floor for thermal storage, it may meet more than 50% of the heating demand. It has been estimated that solar heating of this kind is already worth over $20 million/year in California alone.

These sun tempered homes can overheat from too much solar gain in the winter unless insulated drapes or shutters are used effectively or the collection area is separated from the house interior. The solarium or solar greenhouse is a very attractive method of doing this. It can provide increased collection area at low cost. Operable vents between the greenhouse and the house allow comfort and can be maintained inside the house, even when it is too warm in the greenhouse.

The solar greenhouse is particularly attractive as a retrofit and is being widely used for this purpose. It provides the added benefits of homegrown food, a pretty space, and added living area.

Heat distribution can also be a problem in the sun tempered house. Even when the south rooms are very warm and comfortable the north rooms may be a bit cool. This can be overcome by careful planning. For example, putting bedrooms on the north side reduces the need for heating. Adding skylights or clerestory windows will also allow heating of the center or north side of the house. Clerestory windows are better—because they are less likely to leak and easier to shutter. Elevating the north rooms will also improve heat distribution.

Another problem with this type of house is that it cools down at night or during a period of cloudy weather. These problems can be solved by adding thermal mass to supplement the heat storage in the tile floor and sheetrock and wood of interior walls. The following types of houses all include thermal mass and work much better for heating and cooling than houses without it.

There is little need for a complex formula to design a simple passive house. The Building Code sets the contraints of glazing at about 15-20% of floor area—depending on area in the country. If about two thirds of this or 10% of the floor area is in windows on the south the house will work reasonably well. Most of the other glazing should be on the north side of the house—except as required for fire egress. Insulated drapes or shutters will help control heat loss at night and overheating during the day.

If a solar greenhouse is added it should have its own thermal mass included. The amount of mass can be calculated using the rules of thumb in the next section. If it does not have mass it may be a heater only (with temperatures too hot and too cold for plants). Greenhouses can be crudely sized as follows, using the house example described under conservation.

$$\text{GLAZING AREA IN SQUARE FEET} = \frac{\text{24 hrs. x Building Heat Loss (conduction \& infiltration at design temperature)}}{\text{Solar Heat Gain, December day}}$$

An example using our earlier material:

$$264 = \frac{\text{24 hrs. x (10,300 + 6,912) BTU}}{\text{1562 BTU from sun}}$$

From this figure we can also subtract the south windows of the house that will still be exposed. For example, if we leave about 64 sq. ft. of window exposed, then only about 200 square feet of glazing on the greenhouse is needed, or for this house a greenhouse with a 10' x 20' window. In fact, because the greenhouse would reduce the conductive heat loss and infiltration an even smaller greenhouse would suffice.

B. Standard Frame House with Freestanding Mass Element

A frame house with a freestanding mass element(s) is one of the simplest and most attractive passive solar heating systems for several reasons. First, it can be very economical as it is easily installed in standard construction. Second, it is very effective for heating and cooling. Third, it can be made very attractive or practically invisible if desired. Fourth, it can allow views outside, avoiding the claustrophobic sensation possible with full mass walls. Fifth, glazing control is easy to install and easier to repair compared to that installed behind a mass wall. And finally, mass elements can be retrofit fairly easily to many houses and commercial buildings.

The most commonly used materials for this type of storage are water (in various containers), concrete block, concrete, stone, or adobe. As we have shown, water is more than twice as effective per unit volume as masonry and costs about half as much. It should be considered first in areas where very hard freezes are not common.

The size and placement of the mass element(s) can be very flexible, to complement south windows, skylights, clerestory windows, or a solar greenhouse. However, the mass will work best if placed directly in the sun and exposed to the room. If it is in the same room but not in the sun it is worth only one third as much, and if it is in a room without sun it is only worth one tenth as much.

A general rule of thumb is to use 30-50/lbs. of water equivalent mass per square foot of south glass. The amount of south glass can be determined using the formula we used for solar greenhouse sizing. The mass in the house will prevent overheating and allow much more glass to be used on the south sall.

To continue with our earlier example:

REQUIRED THERMAL MASS
264 SQ. FT GLASS x 50 lbs./SQ.FT. = 13,200 lbs. of water or 212 cu/ft.

We can provide this in a rectangular water tank along with two thirds of the south wall if it averages 4 feet high and is 2 feet thick. In fact, we would probably make sitting benches eighteen inches wide and 18 inches high in front of some of the windows and 6 foot tall sections or even 8 foot tall sections elsewhere.

We have suggested a water tank made from mild steel in a module because it is one of the simplest methods of providing free standing mass. The tank can be easily installed in most new or retrofit situations at a modest cost. Structural bracing will generally be needed on suspended floors.

The tank can be textured and painted to match the walls so it becomes almost invisible, or painted bright colors for a major statement. The window side should be painted flat black or very dark brown or blue. Other containers have also been used for water storage, including drums, culverts, cans, and bottles.

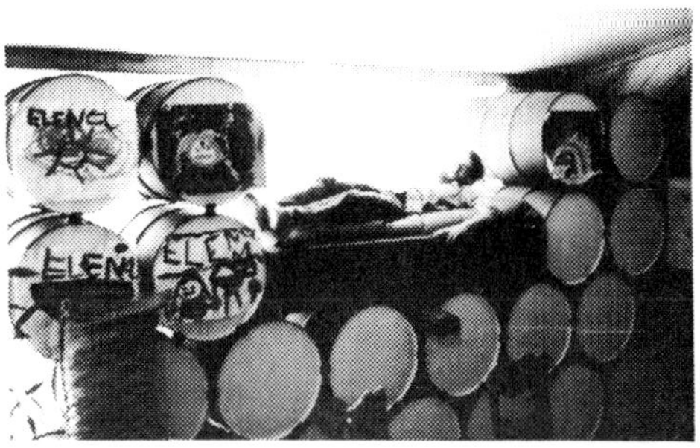

As noted earlier, water is not the only storage element for free standing mass walls. Adobe, stone, concrete, concrete block, and other materials can also be used for heat storage. They will be most attractive for use in areas where freezing of even a full water tank might occur. This is not a common problem in the areas of most rapid growth in the U.S. as it is confined to the north, northeast, and high mountains.

C. Integral Mass Element

When masonry is used for thermal mass storage the volume often becomes so large that a full wall or ceiling is required. The most common type of system using an integral mass element is often referred to as the Trombe wall after the French inventor who has done much of the development of this system. A similar system was patented by Morse in the U.S. in 1882, so it is perhaps more fair to call it the Morse-Trombe wall. These mass elements are usually more expensive than water tanks and are best suited to heating only areas where freezing might be a problem. Vents improve performance significantly and should be fitted with flap valves (made of light plastic) to prevent reverse thermosiphoning. Vent area should be about 1% of total mass wall area.

Integral Mass elements are attractive passive solar retrofits for many older brick and stone houses and commercial buildings. Many of these buildings can be retrofit simply by cutting vents in the wall and adding south glass on the outside of the wall.

To use our simple example once more with masonry instead of water:

Solar Glazing = 264 sq. ft. x 50 lbs water = 13,200 lbs water x 3 = 39,600 lbs masonry at 125 lbs cu/ft = 316 cu/ft of concrete, we could fit this in a 1 foot thick wall with all of the south wall in glass.

The sizing of an integral mass element is similar to that for the freestanding mass. This mass can also be placed in the attic or roof. The disadvantages are obvious - higher cost, more seismic risk, and difficulty. However, this location does save floor space and can be designed to provide more uniform heating throughout the house. The design by Baruch Givoni is representative of this type.

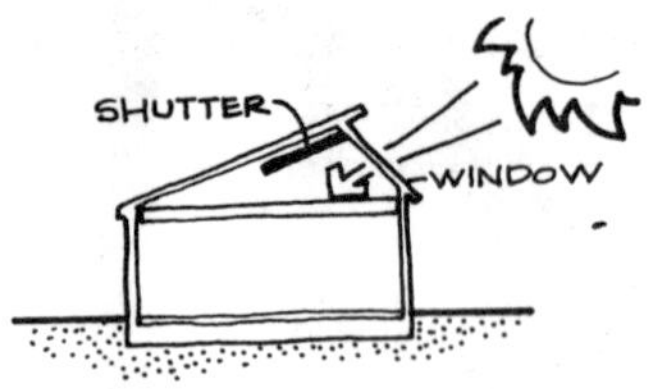

These integral roof elements are more ideally suited to cooling and the various roof ponds will be discussed in the next section.

D. Massive Construction

Thermal mass can also be added throughout the building by using heavy weight construction materials such as concrete block, concrete, stone, or adobe for the structure with insulation on the outside. This construction is typically more expensive than frame construction but it adds thermal mass and with a basic solar tempered design can provide 90+% of the heating in many areas of California. It is also good for cooling. Masonry block and heavy weight construction have a fairly bad reputation for comfort because they have typically not been insulated at all, and when they have been insulated it has been on the inside - the worst place.

More recent work, notably Jesse Savell's method of prefab tilt-up slabs insulated on the outside, has made this type of massive construction much more attractive. The area of glazing can be sized using the technique suggested earlier. Because concrete is slower than water at transferring energy, the use of a solarium or solar greenhouse may be desirable to prevent overheating inside the house. The greenhouse can be placed on the south wall and use the wall as its thermal mass.

Underground or partial underground construction is another form of massive construction. It is also expensive, yet it can provide even better performance for heating and cooling in difficult climates. Rather than sitting up in the environment of wide temperature swings and strong winds the building sits surrounded by the earth at a fairly constant temperate temperature.

This type of building, with its higher cost and more difficult construction is ideally suited to very severe climates or areas with steep hills where site modification must be very extensive anyway.

The size of windows for this type of structure can be calculated using the technique developed earlier. The improved performance of this type of structure is primarily due to the reduced loss coefficient and increased storage. A solar greenhouse can be used to augment heating.

The use of pressure treated wood can considerably reduce the cost of this type of building. Pressure treated wood may also be used to berm existing houses.

3. Backup Heating

In the rapidly growing areas of the U.S. passive solar heating will easily provide more than 50% of the heating demand - in a very conventional house. In a house with more mass and good exposure in these areas 70-80% heating will be relatively easy to achieve. The last 10-20% of the heating in these areas, and in the colder, cloudier, northern areas can be provided most economically with a backup heater.

The most attractive candidate for most people will be an airtight wood stove. These are efficient, require little space and can make you independent of the power system. They can also be purchased or made to include a hot water heater. This can be used either for space heating or hot water. Ken Kern's book **Fireplaces** describes the precautions necessary for this type of system.

A second possibility is the use of a fireplace, but not the typical energy waster. The standard exposed fireplace pumps heat out the chimney flue and walls (R-only 1) and sucks cold air in through windows, doors, and walls. Yet an internal fireplace can work almost as well as an airtight stove if it is designed properly. It should include a heat exchanger, either air or water, damper(s), tight doors, and an outside air intake. These are shown in the accompanying illustration.

Fireplaces, describes many of these modifications in detail and is worth buying. The information it contains may also be applied to an existing fireplace to make it work better. Adding doors is one of the easiest and most cost effective steps. Setting an airtight stove in the hearth will also work. If you live in one of the seismic risk areas shown in Chapter 4 you should consider taking down a fireplace and brick or stone chimney if it is not in good condition and well braced.

Conventional systems should be carefully sized and installed so they operate efficiently and safely. Generally the less frequently the furnace cycles on and off the better it works. This requires sizing it accurately rather than simply letting the installer use a rule of thumb such as "100 BTU capacity per square foot of house." Buy an efficient unit (some of the Swedish oil burners are twice as efficient as the more common furnaces in the U.S.) and keep it tuned up.

Two electrical heaters which look interesting are those that heat water in a radiator and those that heat a radiant panel that looks like a painting. These may be worth a careful look if you cannot get gas or oil service or if you prefer electricity's low pollution (in the house - at the plant is something else again).

There are three other systems which should be considered. The first is an active solar system. It may be viable to backup your passive heating system with an active system and insulated storage tank. It can extend solar heating through additional cloudy periods or provide added heating on very cold, clear days. In the summer it could be used to heat hot water.

The second is more direct and uses an active system to heat up the passive storage directly. This may be most effective in areas where water storage is used but can also be used with masonry walls or floors. The active system is usually very efficient when used in this manner because the operating temperatures are quite low.

The final system worth looking at is a wind heating backup, either mechanical or electrical. In most areas the highest heating loads will occur when the wind is blowing hardest. This can make wind heating very competitive with gas. Wind energy is renewable and independent of the power system. **Harnessing the Wind for Home Energy** by Dermot McGuigan has a good introduction to this use of the wind.

4. Passive Cooling

Solar control and convective, conductive, evaporative and radiant cooling can be combined to provide comfort anywhere in the U.S. Although it may be expensive in some of the hottest areas, costs should at worst be comparable to mechanical systems.

7 Rules For Passive Cooling

i. Make sure the house is well insulated and weather tight.

ii. Make sure the house is oriented properly with most windows on the south and north, few on the east and west.

iii. Provide shading for windows.

iv. Use light colors for roof.

v. Provide windows, vents, and fan if needed for good convective cooling.

vi. Use the simplest, smallest, and cheapest thermal mass needed to store nighttime cooling for use during the day.

vii. Consider heating as well, and integrate design to provide heating and cooling at lowest possible cost.

These methods can be combined in a number of ways. We will include detailed descriptions of only five types:

A) The Standard Frame House
B) Standard House with Free Standing Mass Element
C) Massive House
D) Integral Mass Element
E) The Submarine
F) Other Passive Cooling Methods

These are followed by a look at some of the unexplored options.

A. Standard Frame House

The standard frame house can function fairly effectively as a passively cooled home if it has good solar control and landscaping, uses light exterior colors, and has good natural ventilation at night. In areas with cool night temperatures a house of this type with exposed tile floor may provide full cooling—even with daytime high temperatures near 100°F.

Performance can be improved in areas of marginal nighttime cooling by adding special features to increase natural ventilation or by installing exhaust fans for added ventilation. As the night temperatures get warmer more and more ventilation will be required to achieve a given level of cooling.

This type of house will, at the very least, greatly reduce the energy demand for cooling and will extend full natural cooling further into the summer from spring, and end it sooner in the fall. The most common problem will be lack of carryover or storage of nighttime "coolth" for use during the day. This can be provided by adding thermal mass.

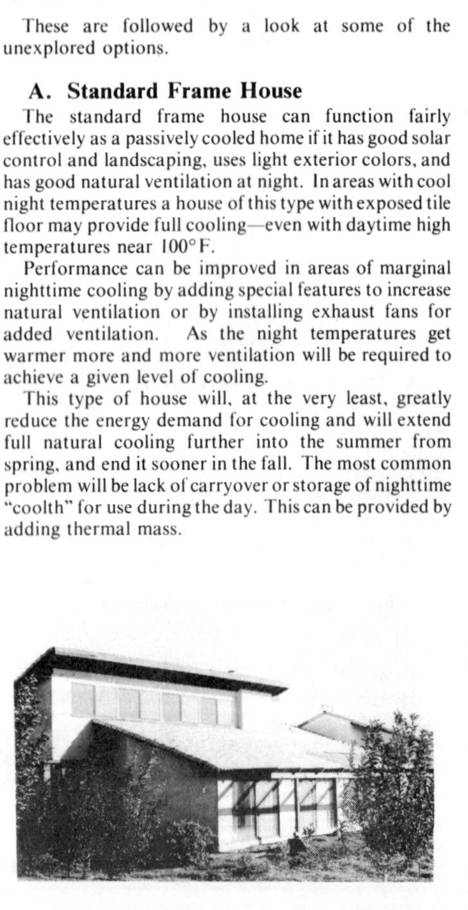

David A. Bainbridge

B. Standard House with Free Standing Mass Element

Adding thermal mass as a free standing mass element can considerably improve the cooling performance of a standard frame house. The use of water storage is particularly attractive because the mass can dissipate heat faster—an important factor if night temperatures are cool for a relatively short time.

The water wall houses in Davis demonstrate how effective these systems can be. With good solar control, adequate thermal mass, and fairly good flow through ventilation convective cooling can meet the full cooling needs of the area—despite daytime highs over 100°F. The accompanying chart illustrates the performance of a waterwall house in Davis, California.

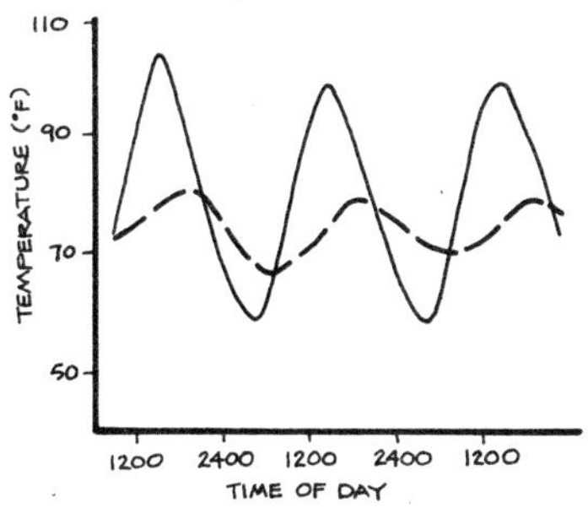

The water line for irrigation can be routed through the tank for additional cooling while irrigating or for emergency core cooling during a worse event.

Cooling with this type of house could also be further improved using pipes set in the earth to draw in cool air. This will work best if the house is set on a hill or bermed (to reduce flooding problems). Vents should be screened carefully. Cooling rates from this type of system are poorly known.

The amount of cooling that can be accomplished with cross ventilation can be calculated:

Vc = air removed (cfm) x .018 x 60 x (Ti - To) (BTU/hr)

For our example house then with a 7 mph north wind: air removed/min = 4250 (outlet/inlet ratio 3:1) x 80 x 7

If the outside temperature is 60° between 12 PM and 5 AM then

Vc = 4000 x .018 x 60 x 10

Vc = 42,000 BTU/hr x 5 hrs = 210,000 BTU/night

This will change the temperature of our water tank about 2°F. If we calculate the cooling load on the building this will tell us if our mass for heating is enough for cooling. Assume the average outside air temperature is 80°F.

Cooling Load = Conductive gain + infiltration gain + radiant gain = (145 x 10° x 24 hrs.) = 34,800 + (.018 x 8000 x 10° x 24 hrs.) = 34,500 + (20 sq. ft. east & west glass x 895 BTU/ft^2) = 17,900 + (40 sq. ft. exposed south glass x 1720) = 68,800 + (40 sq. ft. north glass x 484) = 19,360 BTU/ft^2

Cooling load = 175,000 BTU/day.

This suggests that we don't need any more mass.

C. Massive House

The massive house, and to a lesser extent the mass wall house, are well suited for nighttime convective cooling. Natural ventilation can be augmented using high and low vents or fans as needed. The underground house, as the ultimate massive house, is particularly good for summer cooling. The temperature at about 10' below the surface closely approximates the yearly average temperature. This is an excellent place to start for natural cooling.

D. Integral Mass Element

The most effective cooling can be achieved with the integral mass house. The most effective is the roof pond house. It combines the use of convective, evaporative, and radiative cooling for quite powerful cooling. It can provide full cooling anywhere in the U.S. Several different variants have been developed including Harold Hay's "Sky therm" house with movable insulated lids, Pittenger's "Energy Roof" which moves the water, and the "Cool Pool" with full shade developed by Jon Hammond.

The "Skytherm" house has been most thoroughly tested and has provided full cooling in Phoenix, Atascadero, and near Sacramento. The insulated lids are closed during the day and opened at night for cooling. It can be built in several forms to meet local site or aesthetic concerns. The lids can either slide, flip up, or roll. The house under the roof pond should, of course, begin with good orientation, insulation, solar control, and ventilation.

The expected performance of a Skytherm house in Indio is shown in the accompanying chart.

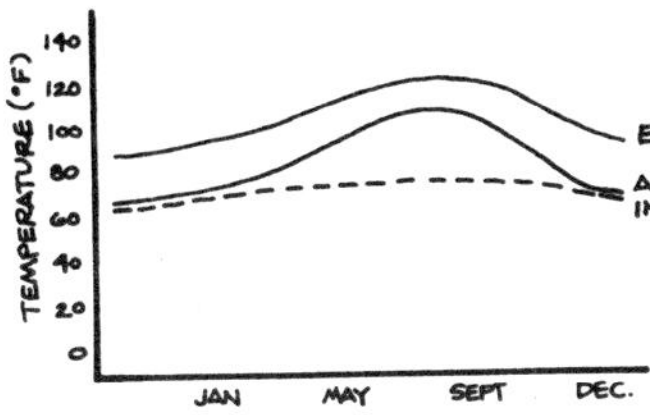

The "cool pool" house is much less complicated. It relies on a fully shaded roof pond for cooling. It requires no movable panels or controls and provides performance similar to the Skytherm house. The cost should be lower than the Skytherm houses. A schematic and test results from Sacramento are shown in the accompanying illustration.

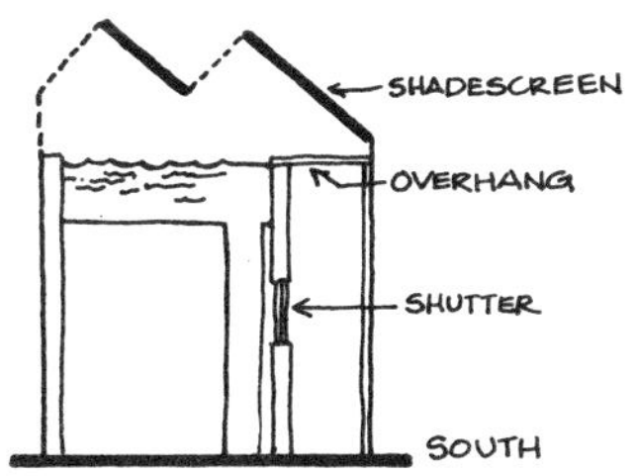

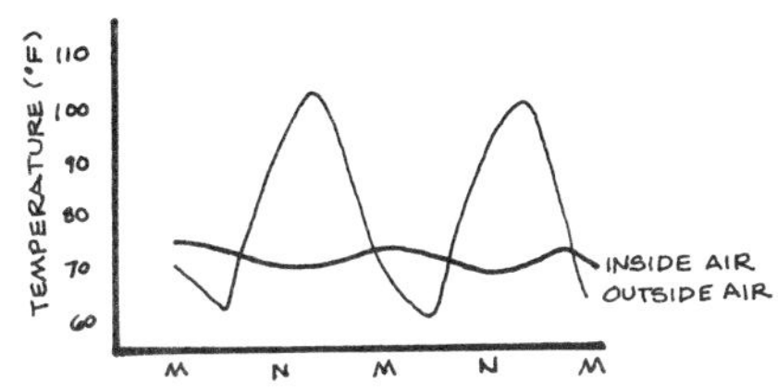

Baruch Givoni has also used integral mass element in the roof but he used concrete instead of water. Reasonably good performance was achieved in the Negev Desert in Israel.

The integral mass wall will also provide cooling with convective cooling. Water walls are best for this type of operation.

E. The Submarine

A final solution is the use of a very lightly massed building. A very simple and elegant solution of this type was developed in the early days of development in the Coachella Valley. It was, and still is, an admirable solution to a difficult climate. The critical period for the early settlers, and even today's residents is at night when sleep can be impossible with "low" temperatures of 95°F. The solution utilizes both radiant, evaporative, and convective cooling. These "submarines," as they came to be called, are lightweight metal sheds with a sprinkler system on the roof. The light shell prevents much storage of the day's heat, the sprinkler provides evaporative cooling, the night sky provides radiant cooling, and the roof provides a very cool radiant surface for the people sleeping inside. The author feels that this idea could also find use in turning an attached solar greenhouse into a cool room for sleeping on hot summer nights.

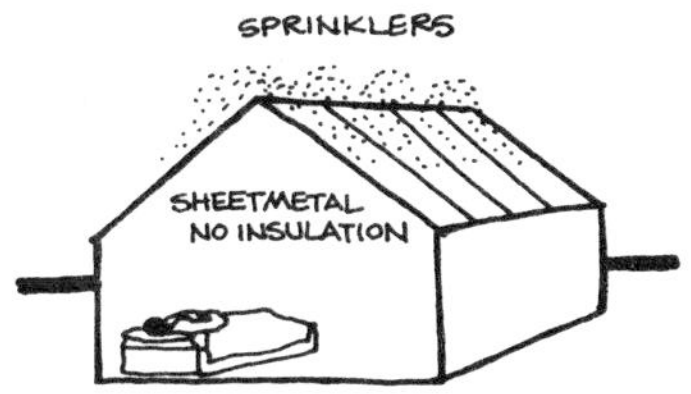

F. Other Passive Cooling Methods

Numerous other designs are possible and should be explored more carefully. These can be tailored for retrofit as well as new construction. The thermosiphon cool slab house below is one possibility. It should work fairly well and cost little more than a standard house. The fluid is cooled in the cool pool and the dense cool water sinks to the bottom. From there it settles into a pipe in the floor slab. As it warms up the lower density fluid circulates up to the top of the cool pool.

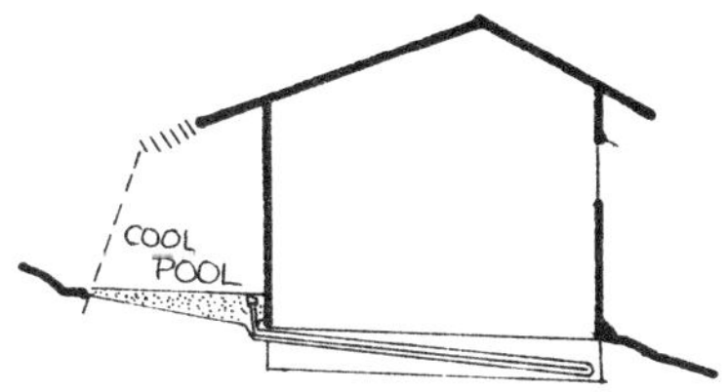

Baruch Givoni has also developed another design which uses solar control and radiant cooling to meet the high cooling demand of the Negev Desert. It includes a central courtyard with inward sloping roofs, and a selectively transparent cover which transmits infared radiation to provide cooling. The selective cover allows the thermal radiation to radiate to space, but prevents convective heat gains from the hot night air. This house should be tested with the more sophisticated material available today. It's cooling could be supplemented with a fountain or pool in the courtyard.

And Frank Lloyd Wright developed an interesting cooling system using a fireplace as a trickle cooler. Water circulated from a pool at the bottom of the chimney to the top and trickled down inside—cooling the mass and setting up a cool air flow into the house.

5. Backup Cooling

Although passive cooling can meet the full cooling demand anywhere in the U.S. a simple passive house or retrofit project will often require additional cooling. Several energy efficient backup cooling systems can be used for added cooling. These range from the casablanca fan to the adsorbtion cycle cooler.

The casablanca fan will probably be the most commonly used backup cooler. By moving the air around a fan will extend the comfort zone considerably. They are in most cases remarkably efficient and will move a large volume of air with only 40 watts. They are also useful in winter as they will bring hot air down from a cathedral ceiling. Some models have a variable pitch to change direction of air flow.

Slightly more complex, yet equally effective is the use of cold ground water for added cooling. This may involve, as in the Davisville house example, running the cold water supply line through a heat exchanger in the water wall. Or it can include a large coil of polybutylene or copper pipe in a plastered wall coupled either to the cold water line or to a cool pool. A small electric water chiller, as used for water fountains, could also be coupled to either the water wall, radiant wall, or to radiant surfaces in a canopy bed.

A more common solution will use a direct evaporative cooler for peak cooling periods. These are available in many sizes, from very large for industry on down to mobile home or auto sizes. These are very effective in dryer areas and provide comfortable cooling at one fourth the energy cost of air conditioning. In humid areas with severe cooling demand they may not work well unless they are used to cool a radiant wall or Trombe wall rather than directly affecting living space.

A more attractive system in these areas is the indirect evaporative cooler. Unfortunately, these will have to be custom made in most areas because their manufacturers were largely driven out of business by subsidized electricity and absorption cycle air conditioners. J.R. Watts, **Evaporative Air Conditioning** has an excellent section on indirect evaporative coolers. A type he does not include but is worth considering is the regenerative rock bed cooler now used widely in Australia.

A more sophisticated and expensive approach is the possible use of an adsorbtion or dessicant cycle evaporative cooler. This involves some type of dessicant material in a heat exchanger coupled to an evaporative cooler. They will be extremely useful when they become available. The dessicant can be dried with an active air solar system.

In some areas, where temperatures are moderate but humidity is higher a dehumidifier may be better than an air conditioner. An extremely efficient air conditioner may be required in some cases. Make sure that the efficiency or EER value is at least 10 and size it very carefully to just meet the demand. They are often sized by rules of thumb which will be completely wrong for an efficient passive house. Install it in a cool place, north is best, and keep it tuned up.

A water heat exchanger may help reduce energy use of the air conditioner. It could be placed in a cool pool, spring, or even in a swimming pool. It could reduce the temperature it is working against from 100°F to only 60°F. This will save energy.

And finally if you are interested in a fully solar house take a look at absorption cycle cooling using active collectors to drive a refrigerator cycle. Prices will come down only if people buy them!

Energy Conservation

Hammond, J., Hunt, M., Cramer, D. and Neubauer, L. (1974) **A Strategy for Energy Conservation**, Living Systems, Rte 1, Box 170, Winters, CA 95694.

Hammond, J., Kopper, B., Bâinbridge, D., Et. al. (1976) **The Davis Energy Conservation Report**, Living Systems, Rte. 1, Box 170, Winters, CA 95694.

Vandervort, Donald, Ed. (1978) **Insulation and Weatherstripping**, Sunset Books, Lane Publ., Menlo Park, CA 94025 $2.95.

Robinson, Steve, and Dubin, Fred. (1978) **The Energy Efficient Home**, The New American Library, P.O.Box 999, Bergenfield, N.J. 07621 $5.70.

Busberry, Peter. (1978) **Building for Energy Conservation**, Halstead Press, John Wiley and Sons, New York, N.Y.

Educational Facilities Lab. (1974) **Energy Conservation and the Building Shell**, E.F.L., Menlo Park, CA.

Dubin, Mindell, Blume Associates (1974) **Energy Conservation Guidelines for Buildings**, General Services Admin., USGP, Washington, D.C.

Stewart Bryne (1974) **The Arkansas Story**, Owens Corning, Toledo, OH.

AIARC (1975) **Energy Conservation Building Design**, AIA, 1735 New York Ave., N.W., Washington, D.C.

Steadman, Philip (1975) **Energy, Environment, and Building**. Cambridge Unive. Press, $5.95.

Stein, Richard (1977) **Architecture and Energy**, Anchor Press/Doubleday, $6.95.

NAHB **Thermal Performance Guidelines**, National Association of Home Builders, 15th and M Sts., NW, Washington, D.C. 20005, $10.00.

Portola Institute (1976) **Energy Primer**, Portola Institute.

Burt, Hill, Kosar, Rittleman Assoc. (1978) **Minimum Energy Dwelling**, Dept. of Energy, SAN/1198-1, NTIS, $9.00.

Building Research Establishment. (1975) **Energy Conservation Buildings**, BRE, Watford, England.

Passive Solar Heating and Cooling

A Starter Set

Antolini, Holly Ed. (1976) **Solar Heating**, Sunset Books, Lane Publ. Co., Menlo Park, CA 94025, $2.95.

Anderson, Bruce and Michael Riordan (1976) **The Solar Home Book**, Chesire Press, Harrisville, N.H. 03450, $7.50.

McCullagh, Jim, Ed. (1978) **The Solar Greenhouse Book**, Rodale Press, 33 East Minor, Emmaus, PA, $8.95.

Van Dresser, Peter. (1977) **Homegrown Sundwellings**, Lightning Tree Press, Santa Fe, N.M. 87501, $5.95.

L.O.F. (1975) **Sun Angle Calculator**, Libby-Owens Ford, 811 Madison Ave., Toledo, OH 43695, $5.00.

Mazria, Ed. (1979) **Passive Solar Energy Patterns**, Rodale Press, 33 East Minor, Emmaus, PA.

Bainbridge, David, Corbett, Judy and Hofacre, John (1979) **Village Homes' Solar House Designs**, Rodale Press, 33 East Minor, Emmaus, PA.

A Serious Set

ASHRAE (1977) **Handbook of Fundamentals**, ASHRAE, 345 East 47th, N.Y., N.Y. 10017, $42.50.

Givoni, Baruch (1969) **Man, Climate, and Architecture**, Applied Science Publishers, London.

Olgyay, Victor (1963) **Design with Climate**, Princeton Univ. Press, Princeton, N.J. 08540, $28.50.

Baer, Steve (1975) **Sunspots**, Zomeworks, P.O. 712, Albuquerque, N.M. 97103, $5.00.

Shurcliff, William (1978) **Solar Heated Buildings of North America: 120 Outstanding Examples**, Brick House Publishing Company, Church Hill, Harrisville, N.H. 03450 PB $8.95, HB $14.95

Wright, Dave (1978) **Natural Solar Architecture**, Van Nostrand Reinhold, NY, N.Y.

E.S.S.A. (1974) **Climate of the U.S.**, U.S.G.P.O. Washington, D.C. or from Cornerstones, 54 Cumberland, Brunswick, ME. 04011, $3.05.

Yanda, Bill and Jim Fisher (1977) **The Food and Heat Producing Solar Greenhouse: Design Construction and Operation**, John Muir Publications.

Barnaby, Caesar, Wilcox, and Nelson (1978) **Solar For Your Present Home**, Publications, California Energy Commission, 1111 Howe Ave., Sacramento, CA. 95825, $2.85.

Hay, H., Haggard, K., and Niles, P. (1975) **Research Evaluation of a System of Natural Air Conditioning**, PB 243498 NTIS, 5285 Port Royal Road, Springfield, VA. 22151, $10.50.

Keller, Merily, Ed. (1977) **Passive Solar Heating and Cooling: Proceeding of the 1st National Passive Conference**, Albuquerque, IA-6637-C, NTIS, $10.50.

Prowler, D. et. al., eds. (1978) **Proceedings of the 2nd National Passive Solar Conference**, Philadelphis, Mid-Atlantic Solar Enrgy Association, 2333 Grays Ferry Ave., Phila., PA 19146, $20.00.

Pearson, Jim (1978) **The Hawaii Home Energy Book**, The University Press of Hawaii, $8.95.

Crowther, Richard (1976) **Sun/Earth**, Crowther Solar Group, 2830 E. Third, Denver, CO 80206, $12.95.

Givoni, Baruch, M. Paciuk, S. Weiser (1976) **Natural Energies for Heating and Cooling of Buildings**, BRS, Technicon, Haifa, Israel.

ASHRAE (1976) **Solar Energy Applications**, ASHRAE, N.Y., $10.00.

Olgyay, V. and Olgyay A. (1954) **Application of Climate Data to House Design**, F.H.A.

Hunt, M., D. Kazama, J. Reiss and D. Bainbridge (1978) **A Simple Solar Home for California**, Energy Commission, 1111 Howe Ave., Sacramento, CA 95825

Smith, Tom and Lee Porter Butler (1978) **The Energy Producing House**, EKOSEA, 573 Mission St., San Francisco, CA 94105, $18.95.

Lee Porter Butler (1978) **EKOSEA Homes: Natural Energy Conserving Homes**, EKOSEA, $26.95.

AIARC (1978) A Survey of Passive Solar Buildings, AIARC 1735 New York Ave. NW, Washington, D.C.

An Attached Solar Greenhouse (1976) The Yandas, Lighting Tree Press, Santa Fe, New Mexico.

A.I.A. (1949) **The House Beautiful Climate Control Project**, A.I.A.

Backup Heating: Wood

Kern, Ken and Steve Magers (1978) **Fireplaces**, Owner-Builder Public., P.O. 550, Oakhurst, CA 93644. $7.00.

Shelton, Jay and Andrew Shapiro (1976) **The Woodburners Encyclopedia**, Vermont Crossroads Press, Waits Field, VT.

Gay, Larry (1974) **Heating with Wood**, Garden Way Publishing, Charlotte, Vt. $3.95.

NFPA Booklets on Firesafety with Wood and Chimneys, 470 Atlantic Ave., Boston, Mass. 02210.

Active Solar

Lucas, Ted (1977) **How to Use Solar Energy**, Ward/Ritchie Press, Pasadena $7.95.

Anderson, B. and S. Oddo (1977) **The Solar Age Catalogue**, Solar Visions, Inc., Hurley, N.Y.

Brooks, F.A. (1936) **Solar Energy and Its Use for Heating Water in California**, U.C. Berkeley, Ag. Experiment Station, Berkeley, CA.

Meinel, A.B. and M.P. (1976) **Applied Solar Energy**, Addison-Wesley, Menlo Park, CA.

Kreider, J.F. and Kreith, F. (1976) **Solar Heating and Cooling**, Hemisphere Publ., Washington, D.C.

Duffie, J.A. and Beckman, W.A. (1974) **Solar Energy Thermal Frames**, John Wiley, M.Y.

Wind

McGuigan, D. (1978) **Harnessing the Wind for Home Energy**, Garen Way, Charlotte, VT 05445.

Pierson, R. (1978) **Technician's and Experimenters Guide to Using Sun, Wind and Water Power**, Parker Publishing Co., N.Y. (excellent).

Hackteman, M. (1975) **The Home-Built Wind Generated Electricity Handbook**, Earthmind-Peace Press, Culver City.

Backup Cooling

Gates, S., Baughn, J., and Rosenfeld, A. (1978) "Energy Conservation Cooling for the Sacramento Area," **2nd National Passive Conference Proceedings**.

Lee, M. (1978) "Evaporative Cooling" Presented at the Helioscience Institute Conference in Palm Springs, McGraw-Edison Company, Phoenix, AZ.

Watt, J.R. (1963) **Evaporative Air Conditioning**, Industrial Press, N.Y.

Cooling

Olgyay, Victor and Aladar (1975) **Solar Control and Shading Devices**, Princeton Univ., $7.50.

Gillespie, John Ed. (1977) **How to Build Patio Roofs**, Sunset Books, Lane Publishing Co., Menlo Park, CA 94025., $2.95.

Chapter 6: Twelve Examples

The twelve examples presented here were chosen from the thousands of existing passive solar buildings to show that passive solar design is simple, economical, and attractive. Four of these building were owner designed and built, four were designed by architects, and four were designed by building designers. They were all built using readily available materials and standard methods. Many cost less to build than a conventional building yet use only half as much energy for heating and cooling.

The twelve examples represent the many different types of passive buildings. Considerably better performance could be achieved by many of these through more careful passive solar design - but they all work well as they are. My favorite example is the Miller House, owner-built in 1950 with rammed earth. It has probably the lowest energy cost for building and suggests what is possible even with very resource conserving design.

The figures used for percent heating and cooling are difficult to verify because only limited monitoring has been done for either passive solar or non-solar houses. The air temperature for the passive house can vary more widely than a non-solar house and still remain in the comfort zone.

A Classic Solar Home - The Miller House 1950

Description

The Miller house is one of the best examples of rammed earth construction in the U.S. It also very clearly reflects the value of incorporating simple passive design in a house with massive walls - even if they are not insulated on the outside. The comfort of the Miller house is attributable in large part to the passive design and massive construction but David and Lydia Miller's delightful personalities also have something to do with it.

The house they now live in was one of several simple passive solar rammed earth houses they built in the late '40's and early '50's. It includes good southern exposure for winter heating. Snow in the backyard in the colder months makes a good reflector to increase heat gain and provide full heating on a sunny day. Even on the coldest days the house may get warm enough so that it is desirable to close the draperies or open a window. Backup heating is by gas heated water pipes in the floor slab, with integral color in the concrete, and a fireplace.

In the summer wide overhangs shade most of the south windows. Morning and afternoon have better exposures with the shallow V plan to provide good heating in the morning and evening in the fall and spring. The massive construction keeps the house cool all summer. High and low vents were provided for added ventilation in the house but are rarely used.

The house cost perhaps a little bit less than a conventional house of equal size when it was built in 1950. It has proven itself over 28 years to be comfortable, economical, and virtually maintenance free. The Millers, David an Attorney, and Lydia, a skilled linguist, have traveled widely and visited rammed earth buildings in France, Poland, Russia, Italy, and Germany. Many of these buildings were over 400 years old. I expect the Miller house may well be equally durable.

Owners Comment:

"There never has been a need for air conditioning in the house. It's 75° in here when it's 90° anywhere else. This type of building made good sense in the resource short years after World War II. It makes even more sense today and for tomorrow!"

For further information contact:
David J. Miller
P.O.Box 1424
Greenley, CO 80631

TYPE: Massive Construction
DESIGN: J. Palmer Boggs
CONSTRUCTION: David and Lydia Miller
LOCATION: Greeley, Colorado
LATITUDE: 40°N
ALTITUDE: 4600 feet
CLIMATE DATA: Heating Degree Days (65°) 6100
Cooling Degree Days (65°) 500
SOLAR RADIATION: Percent Possible Sunshine: 75%
HORIZONTAL SURFACE DECEMBER: 20,000 BTU/ft²
SYSTEM DATA: Area·South Glass: n.a.
BUILDING HEAT LOSS: n.a.

Summer Design Temp: 94°F
Winter Design Temp: -9°F
Mean Temp. January: 32°F
Mean Temp. July: 74°F
Yearly Mean Temp: 53°F
Thermal Storage: n.a.
Equiv. Water Mass/ft² glass: n.a.

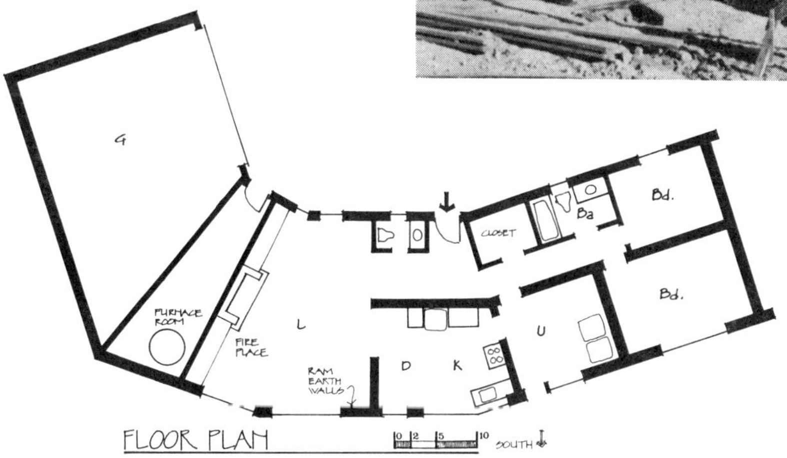

photo: David and Lydia Miller

A Sun Tempered Home - The Schneider House 1977

Description

The Schneider house is about as simple as a passive house can be. It uses an efficient house shell, few east-west windows, good insulation, and careful weatherizing to minimize heating and cooling loads.

Solar heating is provided by three sliding glass doors on the south of the house. Storage is provided in the exposed tile floor used throughout the house. The very open plan enables heat to transfer throughout the house and minimizes overheating. Backup heating is provided by a Franklin wood stove and gas furnace.

Cooling is provided by good solar control and night time ventilation. The open plan and exposed tile floor enables night ventilation to work very well. No backup cooling is needed although on a few afternoons each summer the house air temperature creeps into the low eighties. As the radiant temperatures are still low (and the outside temperature is over 100°F) this still feels comfortable.

Water is currently heated by a conventional gas water heater but the Schneiders are considering installing a flat plate solar water heater. The upstairs area is stubbed out for a dark room.

The house provides full cooling and more than 50% of the annual heating demand. It was owner-built in three months for $17,000. Including the lot the cost was about $25 per square foot.

Owner Comment: "We have really enjoyed the comfort of our house. However, we are now planning a fully passive solar house - with passive hot water and water walls for thermal storage."

For further information contact:
Bob Schneider
2303 Bucklebury
Davis, CA 95616

TYPE: Simple Passive
DESIGN: Bob Schneider
CONSTRUCTION: Bob Schneider
LOCATION: Davis, California LATITUDE: 38.5°N
ALTITUDE: 30 feet
CLIMATE DATA: Heating Degree Days (65°) 2814
Cooling Degree Days (65°) 1063
SOLAR RADIATION: Percent Possible Sunshine 79°
ANNUAL DIRECT + DIFFUSE 90°S: 341,642 BTU/ft²
ANNUAL DIRECT + DIFFUSE 60°S: 552,552 BTU/ft²
DECEMBER DIRECT + DIFFUSE 90°S: 27,370 BTU/ft²
DECEMBER DIRECT + DIFFUSE 60°S: 31,234 BTU/ft²
SYSTEM DATA: Area South Glass: 120 sq. ft.
BUILDING HEAT LOSS: n.a.

Summer Design Temp.: 100°F
Winter Design Temp: 30°F
Mean Temp. January: 45°F
Mean Temp. July: 75°F
Yearly Mean Temp: 60°F
Thermal Storage: 37,000 lbs tile & concrete
Equiv. Water mass/sq. ft. south glass: 40 lbs

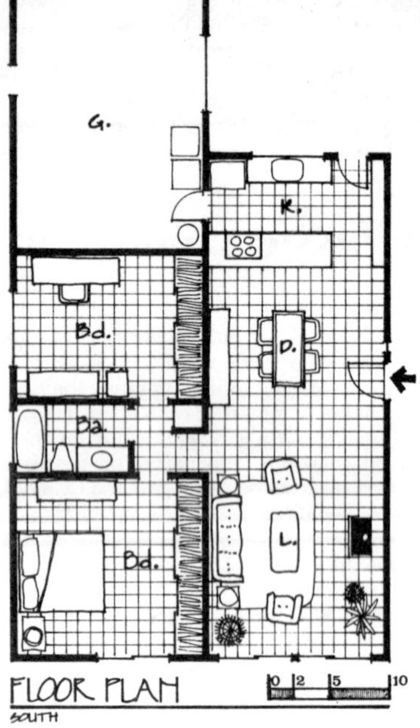

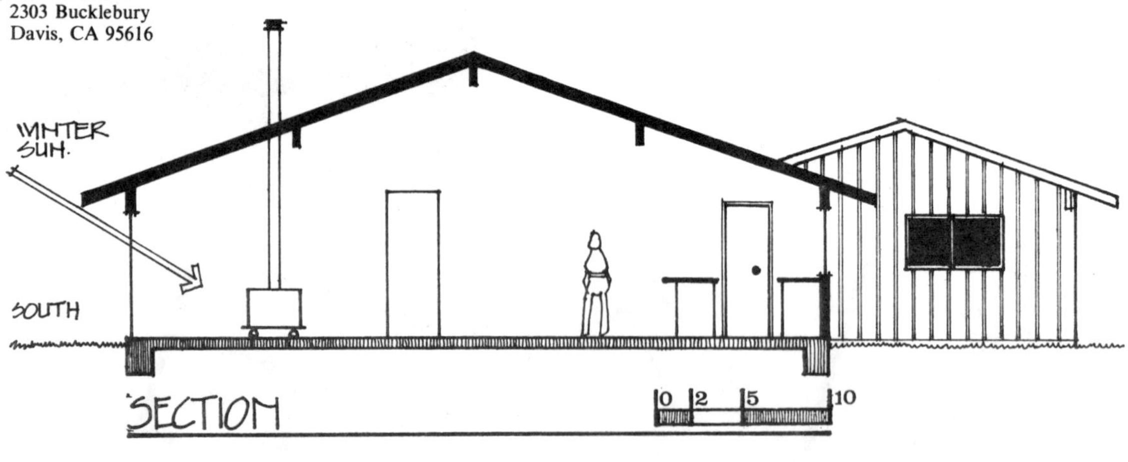

The Bainbridge House - A Simple Water Wall 1978

Description:

This simple 930 sq. ft. house includes 2 bedrooms, bath, dining-living room, and kitchen. It also has a 300 square ft. loft. It uses a rectangular metal tank as free standing mass to augment the performance of a simple passive design.

The house was very carefully insulated with R-19 walls (2"x6" studs) and R-30 roof. Perimeter slab edge insulation is R-5. Insulated shutters and drapes are included on all windows and the skylights. The house was also very carefully weatherized. High quality double pane windows were used to reduce infiltration. Corners and rough framing were also carefully sealed. Interlocking weatherstrip is used on all doors.

Heating is provided by the south glass. Heat is stored in the 2'x12'x4' mild steel water tank, 24 five gallon cans and the exposed concrete slab, which was painted dark brown for better collection. Backup heating is by a Morso airtight wood stove. A small fan and duct are used to take heat from the peak to the north bedroom. The code required gas furnace was used only once - to help dry the sheet rock.

Cooling is provided by night time ventilation of the house. The north/south window orientation is very effective for ventilation in this area. Good solar control was used to reduce the cooling load. South overhangs shade the south windows. West windows were minimized and have shade-screen installed. Light colored walls and roof further reduce heat load. Skylights and loft windows allow ventilation when the breeze is weak. A Casablanca fan is used for backup cooling.

Hot water is provided by a passive solar water heater. It uses three 30 gallon electric water heater tanks for collector/storage. It provides full solar water from March to November and preheating in the winter. A one quart instant hot water tap (180°F) is used to reduce use of the stove and to augment the solar hot water on a cloudy day.

The house cost $43,500 in 1978. The lot cost was $10,500 so the building cost was about $33,000 or $28 sq. ft.

With a very efficient (small) refrigerator utility bills average $5 per month, for about 40 kwh of electricity and 3 therms of gas.

Owners Comment: Dave Bainbridge

"I wouldn't consider living in any other type of house. It is so quiet and comfortable it makes visiting friends and offices that are not passive seem very archaic and uncomfortable."

For further information contact:

David Bainbridge,
2446 Bucklebury, Davis, CA 95616

TYPE: Free Standing Mass Element, Passive Solar Water Heater
DESIGN: Virginia Thigpen, minor revisions and water heater by David Bainbridge
CONSTRUCTION: Virginia Thigpen
LOCATION: Davis, CA
LATITUDE: 38.5
ALTITUDE: 30 feet
CLIMATE DATA: Heating Degree Days (65°) 2814
Cooling Degree days (65°) 1063
SOLAR RADIATION: Percent Possible Sunshine: 79%
ANNUAL DIRECT + DIFFUSE 90°S: 341,642 BTU/ft^2
ANNUAL DIRECT + DIFFUSE 60°S: 552,552 BTU/ft^2
DECEMBER DIRECT + DIFFUSE 90°S: 27,370 BTU/ft^2
DECEMBER DIRECT + DIFFUSE 60°S: 31,234 BTU/ft^2
SYSTEM DATA: Area South Glass 128 ft^2
BUILDING HEAT LOSS: 4 BTU/Sq ft/DD

Summer Design Temp.: 100°F
Winter Design Temp.: 30°F
Mean Temp. January: 45°F
Mean Temp. July: 75°F
Yearly Mean Temp.: 60°F
Thermal Storage: 7,000 lbs/water
35,000 lbs/concrete
Equiv. Water Mass/sq ft./South Glass: 85 lbs

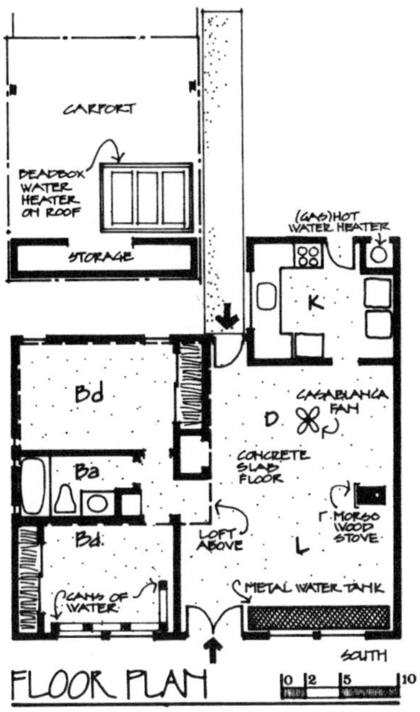

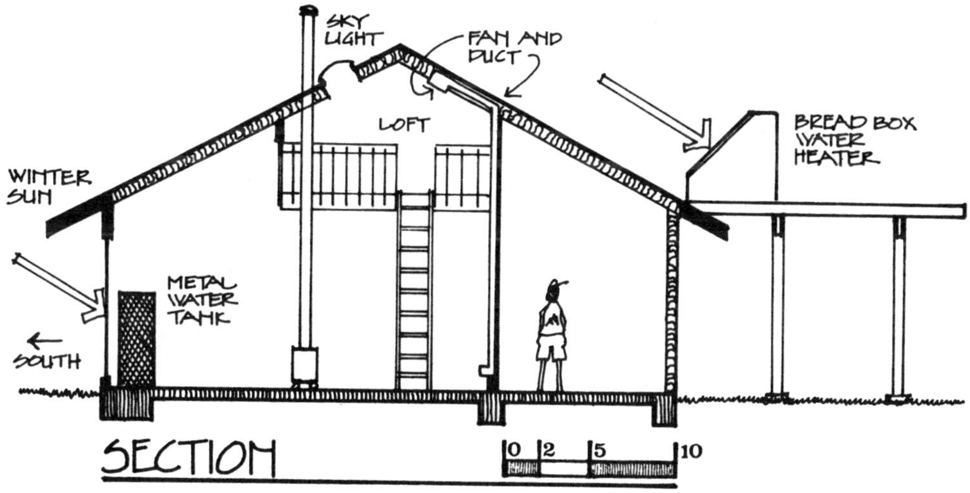

The Davisville House - A Water Wall 1978

Description:

This delightful 1875 sq. ft. house uses its small lot very effectively. The basic shell was made as energy efficient as possible with 2"x6" walls with R-19 insulation and 2"x10" roof rafter with R-30 insulation. Door and window frames were caulked and sealed carefully. Headers are insulated with R-5 foam on the outside. Windows are high quality and seal tightly. Doors have interlocking weather-stripping.

The centerpiece of the house is the 12' high, 17' long, and 1½ foot thick water tank which is built in as the central wall. This stores heat from the sunshine collected by the south windows and skylights. Clerestory windows are used for added heating in the north rooms upstairs. The tank is almost fully exposed on both sides and will provide very comfortable radiant surfaces in four of the rooms of the house. A wood stove and gas furnace are used for backup heating.

The house also cools very well. Night ventilation is very strong using both north/south windows for flow through ventilation and the clerestory windows for thermal stack ventilation. The large exposed area of the tank (almost 400 sq. ft.) will enable it to cool very efficiently. Cooling loads are of course, minimized with good solar control; using shading, light colors, and a shake roof. Cooling can be augmented using a water heat exchanger in the tank that can route all the cold water used for irrigation through the tank.

Hot water is provided by a five tank breadbox heater with glazing set flush in the roof. The two outer tanks on each side are in parallel while the 2 sets of two and final tank are in series. Double glazing is used with a night time insulated shade.

The house construction cost was $68,000 or about $36 sq. ft. The passive heating and cooling system cost $3000. Energy conservation measures totaled about $1000. Deleting air conditioning and reducing furnace size saved about $2,000. This is not a bad price for a house with full natural cooling and 90%+ natural heating.

Owners Comment: "This design is a statement of my desire to integrate a good functional design with an effective passive solar design which provides space heating and cooling."

For further information contact:
Paul Fellers
800 South Third
Bozeman, MT 59715
406-587-2788

TYPE: Integral Mass Element
DESIGN: Paul Fellers
CONSTRUCTION: Virginia Thigpen
LOCATION: Davis, California
LATITUDE: 38.5°N
ALTITUDE: 30 feet
CLIMATE DATA: Heating Degree Days (65°) 2814
Cooling Degree Days (65°) 1063
SOLAR RADIATION: Percent Possible Sunshine 79%
ANNUAL DIRECT + DIFFUSE 90°S: 341,642 BTU/ft²
ANNUAL DIRECT + DIFFUSE 60°S: 552,552 BTU/ft²
DECEMBER DIRECT + DIFFUSE 90°S: 27,370 BTU/ft²
DECEMBER DIRECT + DIFFUSE 60°S: 31,234 BTU/ft²
SYSTEM DATA: Area South Glass: 334 ft²
BUILDING HEAT LOSS COEFF. BTU/FT²/DD: 2.64

Summer Design Temp: 100°F
Winter Design Temp: 30°F
Mean Temp. January: 45°F
Mean Temp. July: 75°F
Yearly Mean Temp: 60°F
Thermal Storage Water: 19,000 lbs
Floor Slab: 51,500 lbs
Equiv. Water mass/sq. ft. Glass: 100 lbs

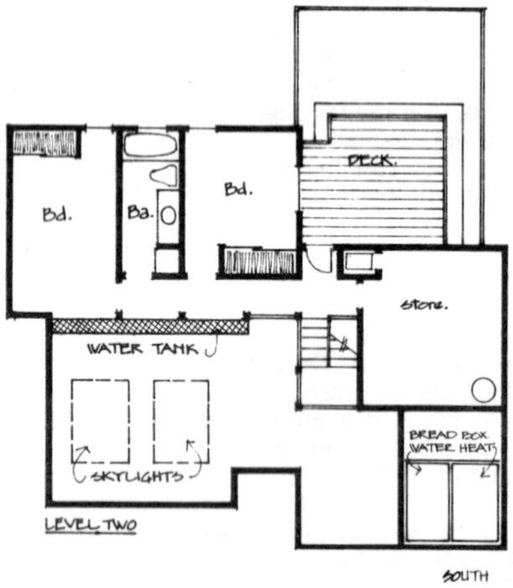

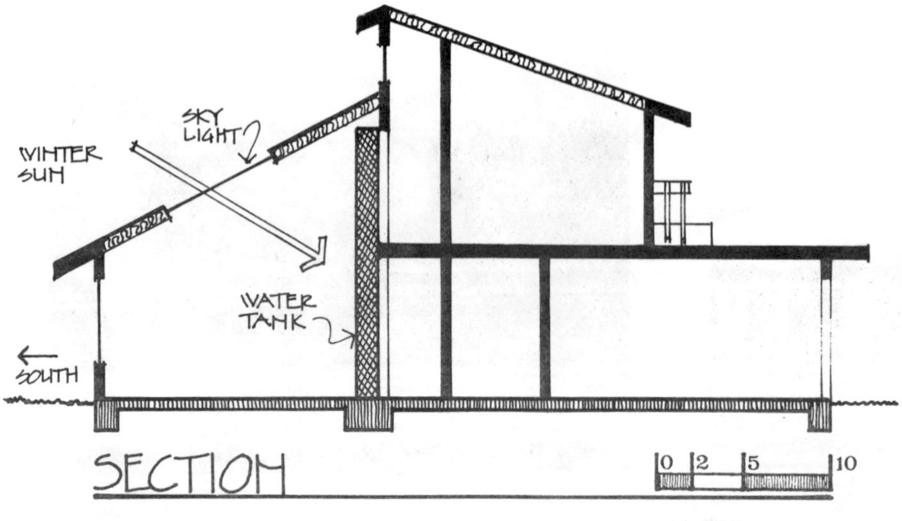

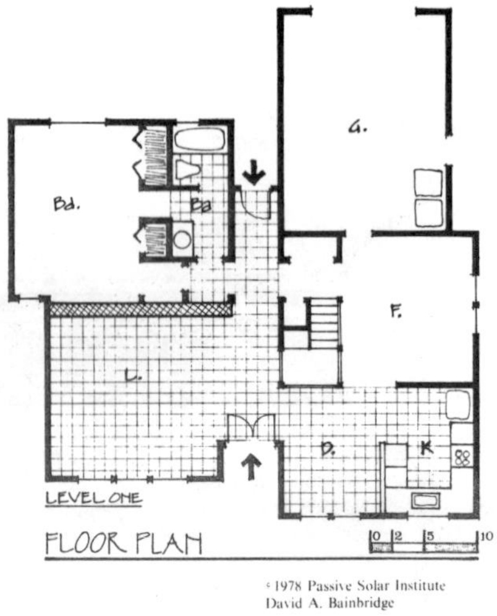

A Solar House for Lake Tahoe - The Smith House 1978

Description:

The Smith house is a very economical and effective 1800 sq. ft. house. It uses the house within a house concept to minimize heating demand in a cold climate. The large greenhouse on the south is very attractive and fits in well at Lake Tahoe.

Heating is provided by the solar greenhouse. Heat is stored in the concrete foundation and interior mass (sheetrock, etc.) and to a limited extent in the backfill under the house. The greatly reduced load of the house-within-a-house makes the solar heating sufficient even in this cold climate. Air is heated in the solar greenhouse, rises to the peak and then settles down the north roof and returns under the house to the greenhouse. The house within is heated by south windows. Only one airtight wood stove is used for backup heating. Insulated shutters or drapes would further improve the house performance.

The house achieves cooling by air circulation through buried pipes, windows in the greenhouse, and the high vents. In a hotter climate an exterior shade would also be necessary.

The house cost has been only $51,800 so far for 1800 sq. ft., although the fully finished cost will reach about $60,000. This is very reasonable and is below average cost in the area. It currently achieves about 80% heating and full cooling.

Owner Comment: Tom Smith

"We have created a climate of warmth and where would you rather be, in a heated room? or in Hawaii?"

For further information contact:

Tom Smith, P.O. Box 2356, Olympic Valley, CA 95730

TYPE: Solar Green House
DESIGN: Lee Porter Butler, John Hofacre, Tom Smith, Bruce Maeda, Carl Gustafson
CONSTRUCTION: Robert Charbonneau
LOCATION: Tahoe City, California
LATITUDE: 39°
ALTITUDE: 6300 ft.
CLIMATE DATA: Heating Degree Days: 8162
Cooling Degree Days: 32
SOLAR RADIATION: Percent Possible Sunshine: 60 [RENO]
ANNUAL DIRECT + DIFFUSE 90°S: 368,046 BTU/ft^2
ANNUAL DIRECT + DIFFUSE 60°S: 568,008 BTU/ft^2
DECEMBER DIRECT + DIFFUSE 90°S: 34,132 BTU/ft^2
DECEMBER DIRECT + DIFFUSE 60°S: 38,318 BTU/ft^2
SYSTEM DATA: Area South Glass 395 ft^2

Summer Design Temp. 80°F
Winter Design Temp. -5°F
Mean Temp. July: 61°F
Mean Temp. Jan.: 28°F
Yearly Mean Temp.: 42°
Thermal Storage: Concrete: 544 ft^3
Backfill: 2160 ft^3
lbs Water Equiv./sq. ft.: 40 lbs
Building Heat Loss Coeff. BTU/DD/Sq ft: n.a.

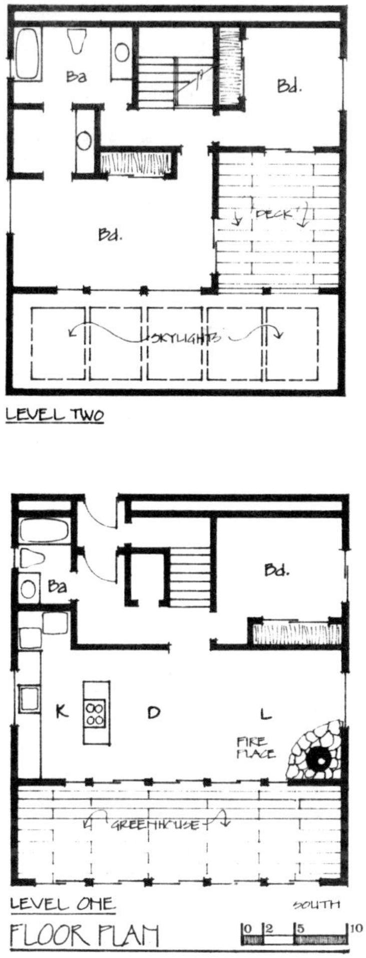

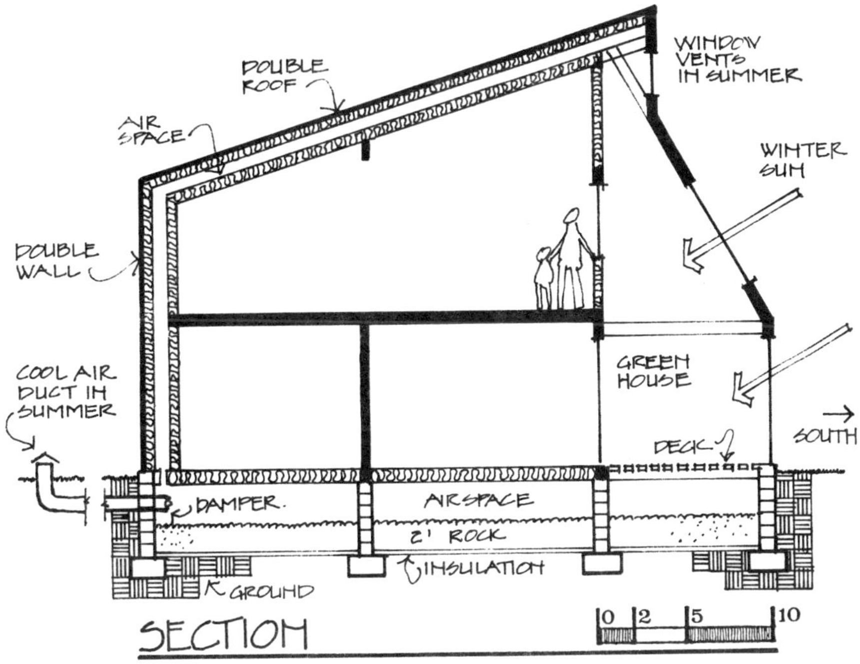

courtesy Tom Smith

Earth Sheltered Living - The Belshe House *1977*

Description

The Belshe house is an attractive 2700 sq. ft. earth sheltered passive house. The owner/builder/designers are justifiably pleased with its beauty and performance. It is an excellent response to the very cold winters as it has very low infiltration and conductive losses. R-5 insulated shutters are installed at night on the double glazes south windows.

The winter heating is provided by the south windows. The small glazing area (only 200 sq. ft.) is acceptable because the load is greatly reduced and the winter sun is not exceptionally reliable. However, when it does shine the snow fields act as a good reflector and very good heating is possible. A heat circulating fireplace is the major backup with electric heating if necessary.

Summer cooling is by solar control, convective, and conductive cooling. The relatively low ground temperature (50°F) ensures pleasant temperatures all summer.

The open plan makes day lighting work even without skylights in the roof. This pleasing natural light removes any apprehension about living underground. We will all become more familiar with this as type of building in the future.

TYPE: Underground Simple Passive
DESIGN: Joe and Mary Belshe
CONSTRUCTION: Joe and Mary Belshe
LOCATION: St. Cloud, Minnesota
LATITUDE: 45°
ALTITUDE: 1034 feet
CLIMATE DATA: Heating Degree Days (65°) 9000
Cooling Design Days (65°) n.a.
SOLAR RADIATION: Percent Possible Sunshine: 45%
ANNUAL HORIZ. SURF: 1352 BTU/Sq. Ft./Day
DECEM. HORIZ. SURF. 561 BTU/Sq. Ft./Day
SYSTEM DATA: Area Glass: 200 Ft.²
BUILDING HEAT LOSS COEFFICIENT: n.a

Summer Design Temperature: 90°F
Winter Design Temp: -20°F
Thermal Storage: Concrete, Slab, Walls, and Berm

For further info:
Joe Belshe
P.O. 1512
St. Cloud, Mn. 56301

Susan Maas Paul

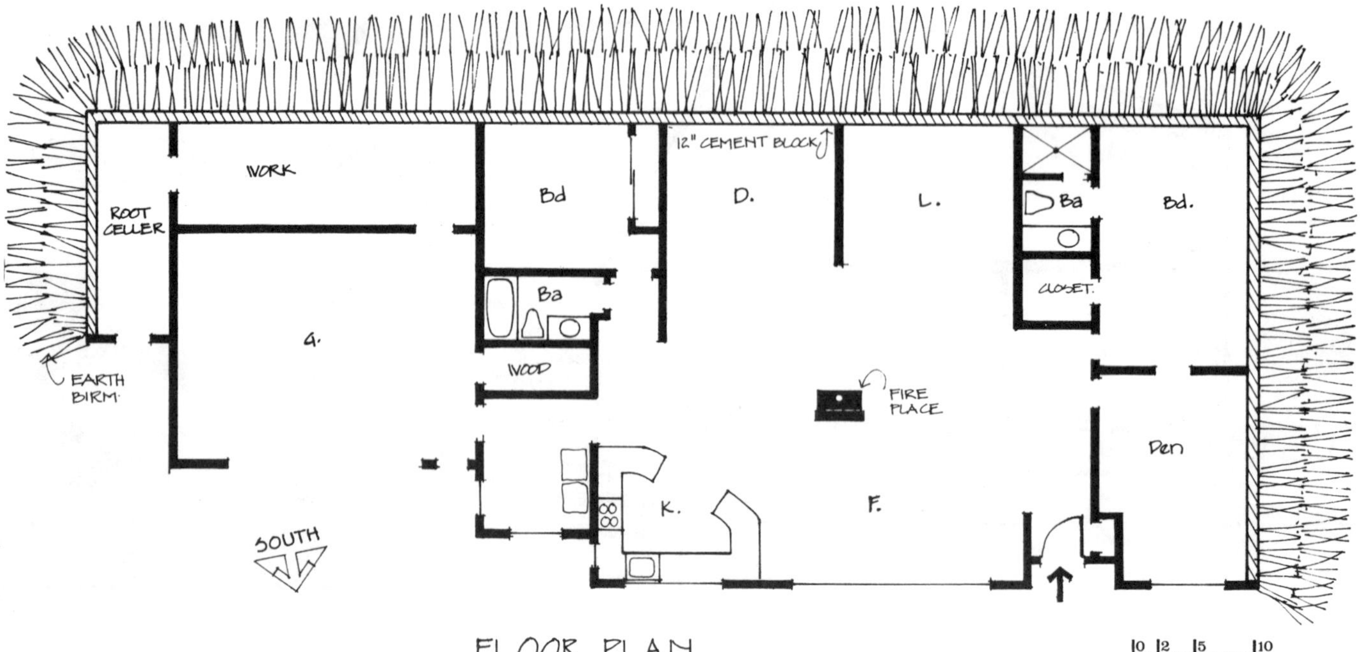

FLOOR PLAN

A Proven Roof Pond - The Winters House 1975

Description:

The Winters House has worked very well and clearly supports Harold Hay's claims for the "Sky therm" system. This very simple and elegant 1,200 sq. ft. house has been in operation for three years now. The roof pond adds extra cooling power to this simple passive house. It is needed on this site due to partial blockage of the cool summer night breeze by the surrounding hills.

The house has 2"x4" walls with R-11 insulation. The windows all have insulated shutters. The key elements of the house are the six 6'x8' galvanized metal tanks in the roof. These are tar lined and include a water-filled plastic bag about 10" deep.

An insulated hinged lid covers the ponds. It is lifted by hydraulic lifters activated by a small electric pump or bicycle drive. The south facing underside is lined with reflective foil.

Winter heating relies on both the south windows and roof pond. Heat is stored in the exposed slab, water filled drums, and the roof pond. The insulated shutters on the windows and the roof ponds are closed at night to reduce heat loss. Backup heating is provided by wood stoves fueled with wood from the owners orchard.

Summer cooling is provided by solar control, with good overhangs and arbors; convective cooling, with night ventilation; and radiant cooling from the roof ponds. It has been necessary to flood the roof ponds for evaporative cooling.

The performance is estimated at 90% heating and full cooling. The house cost $30,000 to build in 1975.

For more information contact:
Jon Hammond, Living Systems
Route 1, Box 170, Winters, CA 95694

TYPE: Roof Pond with insulating lids
DESIGN: Jon Hammond
CONSTRUCTION: Jon Hammond, solar system; Mark Harrington, house
LOCATION: Winters, California
LATITUDE: 38.5°
ALTITUDE: 100 feet
CLIMATE DATA: Heating Degree Days (65°) 2819. Cooling Degree Days (65°) 1063
SOLAR RADIATION: Percent Possible Sunshine 79%
ANNUAL DIRECT + DIFFUSE 90°S: 341,645 BTU/ft²
ANNUAL DIRECT + DIFFUSE 60°S: 552,552 BTU/ft²
DECEMBER DIRECT + DIFFUSE 90°S: 27,370 BTU/ft²
DECEMBER DIRECT + DIFFUSE 60°S: 31,234 BTU/ft²
SYSTEM DATA: Area South Glass: 160 ft²
Area Roof Pond Exposed: 288 ft²
Effective Reflector Area: 120 ft²
Equiv. Water Storage Mass: 60 lbs/ft²

Summer Design Temp: 100°F
Winter Design Temp: 30°F
Mean Temp. January: 45°F
Mean Temp. July: 75°F
Yearly Mean Temp.: 60°F
Thermal Storage: 16,000 lbs water
40,000 lbs concrete

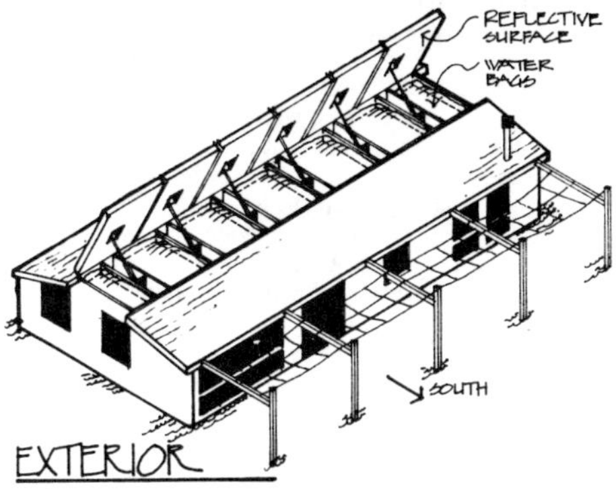

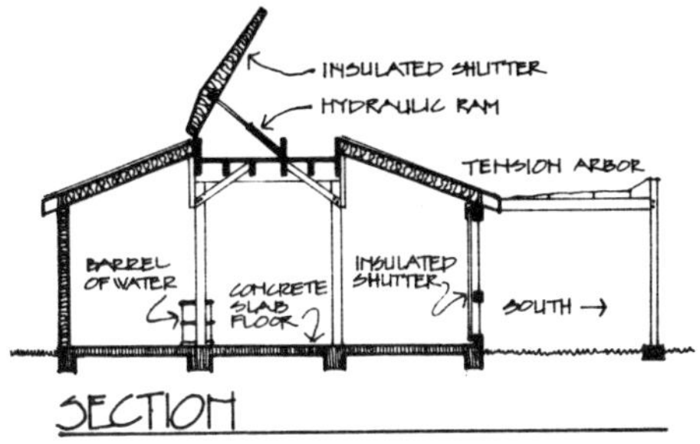

A Classic Retrofit - The Reiss House 1978

Description:

The Reiss House is a 1,300 sq. ft. Spanish style house built in 1932. It sits in one of the older residential areas of Sacramento. The house is on the south side of an east-west street lined with large trees. The south facing back yard was one of the features of the house that Jeff was looking for when he was house hunting.

As you approach the house there is no clue that it is different from the other houses in the neighborhood. However, as you walk through the house to the kitchen your attention is caught by a stained glass sun in the window and drawn to the greenhouse in the back of the house.

The spacious (21' x 7½') greenhouse is very nicely detailed and looks attractive from both inside the house and from the backyard. The fact that it faced about 30° west of south (due to the orientation of the house) would be undesirable in some areas but the morning fogs in the winter actually make it a good orientation for this area.

The double pane glazing (almost 200 sq. ft.) allows the sun to shine on ten 55 gallon drums full of water and the exposed floor. The heat is transferred to the house through vents and existing windows. Since the house is built on three levels with almost all of it above the green house a strong natural convection loop is established which can provide heating to most of the house.

The greenhouse should provide at least 50% of the house's heating needs. As the efficiency of the house is improved with better window treatment and more insulation it should reach 70%. In the summer the green house is almost fully shaded by trees and can be vented directly outside. The cool summer night breezes will enable the greenhouse to provide some cooling for the house. In addition, thermal chimneys will be installed to draw hot air out of the upstairs and pull cool air in from under the house.

The main greenhouse structure was built during a weekend Habitat Center "hands on" workshop directed by Lynn Nelson and Paul Bunker. Jeff and John did much of the preparation and finishing work. The materials cost about $1600 or about $8 per square foot.

The hot water is provided by a passive solar water heater. Because there wasn't a particularly good existing location for it, Jeff, John and friends built a strong roof deck to shelter the back door and support the heater. The water heater uses two recycled 40 gallon water tanks connected in series. A reflector is used to increase winter heat gain.

The passive solar water heater is double glazed with Kalwall fiberglass to reduce night time heat loss. It will be fitted with a roll down insulated cover as time allows. Total materials cost of the breadbox was about $400. It will provide full hot water for about 9 months of the year.

Owner Comment: Jeff Reiss

"I think my house clearly demonstrates the practicality of retrofitting passive design features on existing homes. The passive features have increased the comfort, beauty, and value of the house and reduced the energy bills. Who could ask for more than that!"

For more information contact:
Jeff Reiss
4438 G Street
Sacramento, CA 95819

TYPE: Solar Greenhouse, Induced Ventilation and Passive Solar Water Heater
DESIGN: Lynn Nelson of the Habitat Center with Jeff Reiss and John Burton
LOCATION: Sacramento, California
CONSTRUCTION: Habitat Center workshop including Jeff Reiss, John Burton, Lynn Nelson and 15 participants from the community.
CLIMATE DATA: Heating Degree Days (65°) 2843; Cooling Degree Days (65°) 1159
SOLAR RADIATION: Percent Possible Sun 70%
ANNUAL DIRECT + DIFFUSE 90°S: 341,642 BTU/ft^2
ANNUAL DIRECT + DIFFUSE 60°S: 552,552 BTU/ft^2
DECEMBER DIRECT + DIFFUSE 90°S: 27,370 BTU/ft^2
DECEMBER DIRECT + DIFFUSE 60°S: 31,234 BTU/ft^2
SYSTEM DATA: Greenhouse south glazing: 180 sq. ft.
Cooling Design Temp: 100°F
Winter Design Temp: 30°
Mean Temp January: 45°
Mean Temp. July: 75°F
Yearly Mean Temp: 60°F
Thermal Storage: 5000 lbs/water
4000 lbs/concrete
Building Heat Loss Coefficient BTU/SQ/FT DD 17
Passive water heater south glazing: 32 sq. ft., 2 40 gallon tanks.
Equiv. water mass/ Glass: 40 lbs.

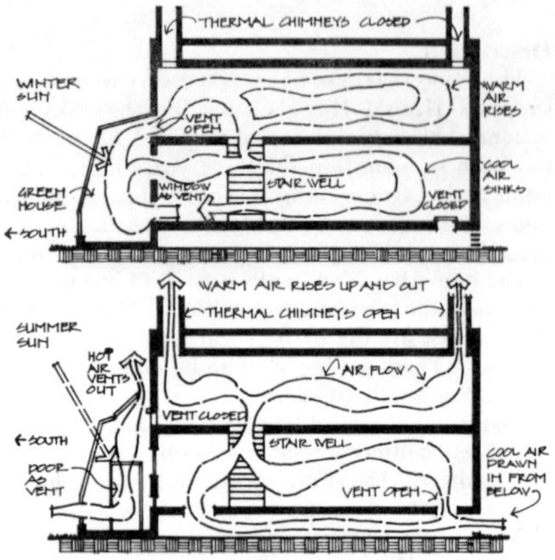

Living With The Sun - Chatauqua Apartments 1977

Description:

This large 108,000 sq. ft., 138 unit apartment complex illustrates the ease with which simple passive design can be incorporated in multi-unit dwellings. It also very clearly shows the value and beauty of designing and living with the sun.

The building shell is standard 2"x4" studwall and stucco. Windows are all on the south and north for good winter heating and summer cooling. The average unit has about 80 sq. ft. of south glass in clerestory and bedroom or living room windows. Thermal storage is limited to interior walls and ceilings as the apartments are almost fully carpeted.

Summer cooling is minimized by use of light exterior colors, shading and landscaping. East and west walls are windowless. South windows are protected by a variety of shading devices, including overhangs, wing walls, and shade structures. Landscaping is used to provide additional protection for the east and west walls. The north-south windows allow excellent night ventilation. The clerestory windows set up good ventilation even when the breeze is relatively weak.

The apartment construction cost of $20/sq. ft. was typical for the area as no unusual features were added. Standard materials and methods were simply used well. The apartments have not been monitored but available data from similar apartments suggests that energy use for space heating and cooling will be about ½ that of similar apartments in the area which do not include passive solar design.

Occupant comment:

"I love the excellent cross ventilation on summer nights. It enables me to stay comfortable with only occasional use of my air conditioner."

For further information:
Bill Rowe
Tandem Properties, Inc.
709 Barcelona Ave.
Davis, CA 95616

TYPE: Simple Passive
DESIGN: John Gustin, Tandem
CONSTRUCTION: Tandem Properties
LOCATION: Davis, CA LATITUDE 38.5°
ALTITUDE: 30 feet
CLIMATE DATA: Heating Degree Days (65°) 2814
Cooling Degree Days (65°) 1063
SOLAR RADIATION: Percent Possible Sunshine 79%
ANNUAL DIRECT + DIFFUSE 90°S 341,642 BTU/ft²
ANNUAL DIRECT + DIFFUSE 60°S 552,552 BTU/ft²
DECEMBER DIRECT + DIFFUSE 90°S 27,370 BTU/ft²
DECEMBER DIRECT + DIFFUSE 60°S 31,234 BTU/ft²
SYSTEM DATA: (Typical Unit) Area South Glass: 80 sq. ft.
BUILDING HEAT LOSS COEFF: n.a.
EQUIV. WATER MASS/GLASS: 20 lbs

Summer Design Temp: 100°F
Winter Design Temp: 30°F
Mean Temp. January: 45°F
Mean Temp. July: 75°F
Yearly Mean Temp: 60°F
Thermal Storage: sheetrock

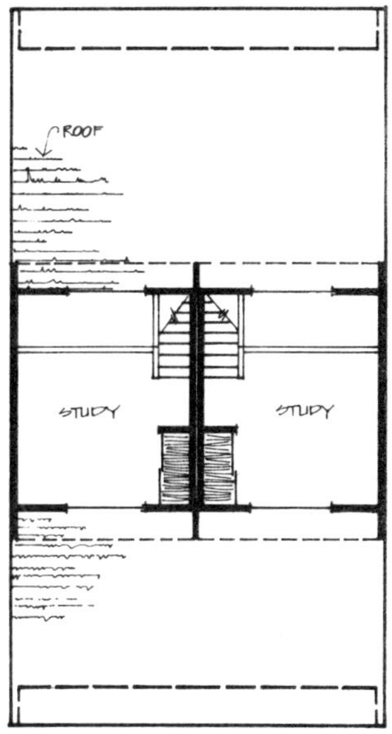

LEVEL TWO

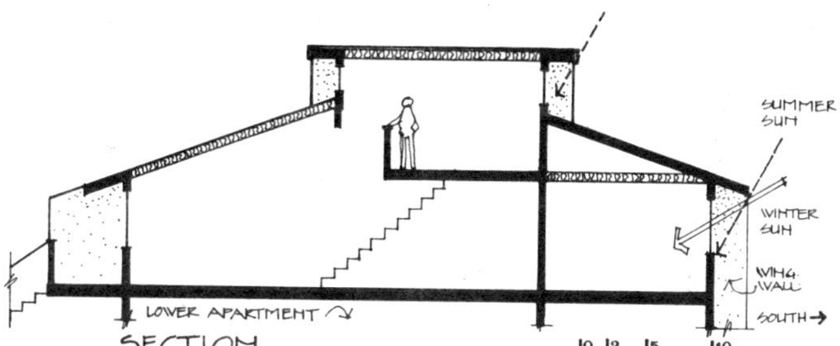

SECTION

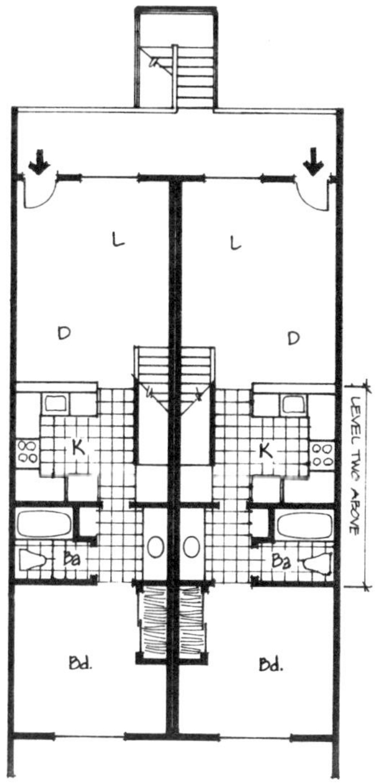

LEVEL ONE

FLOOR PLAN

A Free Standing Mass Commercial Building - The Living Systems Office 1974

Description:

The Living Systems Office is a simple one story building (25' x 40') with the long dimension facing directly south. It has basically four rooms and meets the needs of the architecture, research and design staff very nicely. The outer walls are tilt up concrete slab and wood frame. It is bermed on the east and parts of the north and south. There are no windows on the east or west.

The solar heating is provided by the 250 square feet of south glazing, which is largely double glazed. The sun's energy is stored in steel culverts full of water. There are 7 24" culverts 8 feet high, and 6 12" culverts 3 feet high, for a total water mass of about 20,000 lbs.

The concrete walls and floor provides additional storage. The system provides about 90%+ of the heating. Backup heating is provided by a stove Jon Hammond designed and built from an old submarine net buoy. It features a direct cold air intake.

Full cooling is provided by good solar control and night time ventilation. Louvred shades and wire arbors with vines cover the south windows. Night time cooling is provided by either natural cross ventilation or an exhaust fan and vents. A large tree to the west further reduces building heat gain. Supplemental cooling is provided by a Casablanca fan.

Daylighting is used very effectively throughout the office, with task lighting provided but rarely used. This reduces heat gain in the summer and provides a much more pleasant working environment. The backup lighting is very nicely done with indirect fluorescents. Total lighting is about ¾ watt/sq. ft.

The total energy use for the building is remarkably low for an office of from 5-10 staff. Yet, despite this low energy use (or really because of it) satisfaction of the employees is very high. The building is comfortable, quiet, aesthetically pleasing, and conducive to efficient work.

I worked there for two years and still remember one of the highlights. A reporter was out interviewing us on a very hot summer day concerning one of our present jobs. He seemed more than a little skeptical about passive design but was interested in learning more about it. In the midst of the conversation we observed that the power had gone off (the Casablanca fan stopped turning). We mentioned it to him and asked him to notice the change in comfort and working conditions (NONE!). Suddenly, he seemed to understand what we were talking about. A similar outage in a standard building would send everyone home - from a dark and very hot building.

For more information contact:
Jon Hammond, Living Systems, Route 1, Box 170, Winters, CA 95694

TYPE: Free Standing Mass, Water Wall
DESIGN: Jon Hammond, Living Systems
CONSTRUCTION: Jon Hammond and Friends
LOCATION: Winters, California, LATITUDE: 38.5°, ALTITUDE: 80 feet
CLIMATE DATA: Heating Degree Days (65°) 2819
Cooling Degree Days (65°) 1063
SOLAR RADIATION: Percent Possible Sunshine: 79%
ANNUAL DIRECT + DIFFUSE 90°S: 341,642 BTU/Ft^2
ANNUAL DIRECT + DIFFUSE 60°F: 552,552 BTU/Ft^2
DECEMBER DIRECT + DIFFUSE 90°S: 27,370 BTU/ft^2
DECEMBER DIRECT + DIFFUSE 60°S: 31,234 BTU/ft^2
SYSTEM DATA: Area South Glass: 250 ft^2

Summer Design Temp: 100°F
Winter Design Temp: 30°F
Mean Temp. January: 45°F
Mean Temp. July: 75°F
Yearly Mean Temp: 60°F
Thermal Storage: 20,000 lbs water
Water Storage/Mass sq. ft. glass: 80 lbs/ft^2

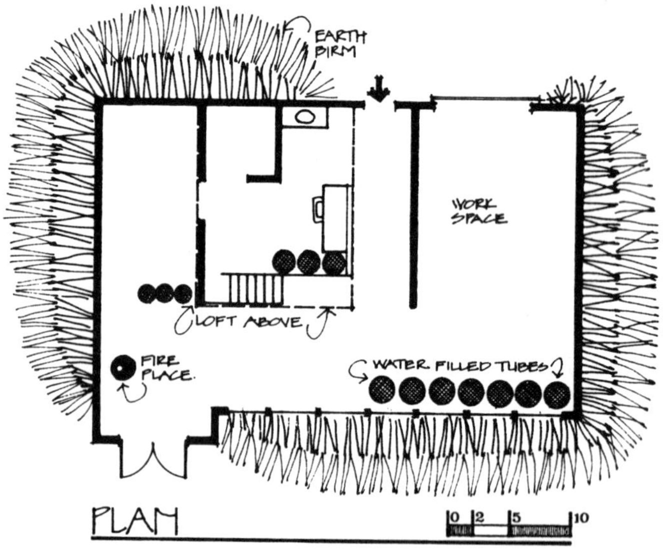

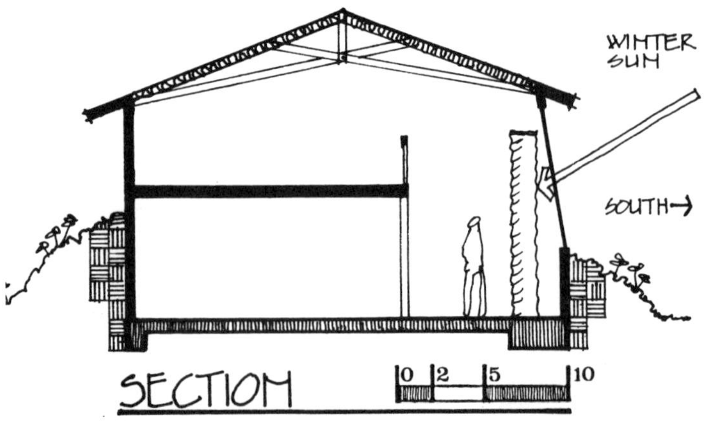

A Commercial Success - The Renault and Handley Buildings

1978

Description

This project includes two 50,000 square foot industrial buildings. They are single story tilt up construction and were built speculatively for lease. In many ways this makes the solar design more difficult because added flexibility must be designed in to suit a wide variety of uses and interior zoning.

The basic design was made energy conserving to reduce heating and cooling loads. Window area is relatively small and double glazed. Insulation was carefully specified and done. Landscaped berms were added where possible to further reduce load. Roof and all exterior colors are light to reduce overheating. East and west glass uses solar control films (shade screen would have been better) and a six foot overhang was added to shade the south wall in the summer.

The passive system is remarkably straightforward and included simple changes in the tilt up wall to make it a passive solar heater. Needed vents were formed in the top and bottom of the wall panels during forming. Thickness was the same as other walls, 5.5", both for simplicity and because the use of the building will be primarily during the day so no long lag is needed.

The wall is painted black on the south side. The south mass wall is double glazed with low iron tempered glass. This is fastened 2" from the wall with standard fasteners and flashing. Perforated metal strips were added to control lateral air flow.

The return air from the system is drawn into the southwall through registers. It rises up the south wall and is collected by a plenum running along the upper vents and carried along by the south wall distributing system. It is backed up by an active solar heated water heat exchanger and auxilary gas heater.

The building area not serviced by the south mass wall collector is heated by an active solar system with 5900 square feet of collector. The Revere collectors are used to heat water in a storage tank. And a heat exchanger is used to supply energy to water coils in the air distribution system.

Cooling is minimized by careful solar control with overhangs and light colors. Back up cooling is mechanical. Provisions have been made to include a solar driven absorption cycle chiller.

The cost of the mass wall system on the first building was about $22,000 for an installed cost of $18.33 per effective square foot of south wall. The active system cost was $130,000 or about $44.80 per square foot. The combined system should provide 90% of the heating demand.

This building is a good example of what can be done using standard building techniques - and some imagination and skill in solar design.

For further information contact:
David Bagshaw, Pacific Sun, Inc.
520 Santa Cruz Avenue, Menlo Park, CA 94025 (415) 328-4588

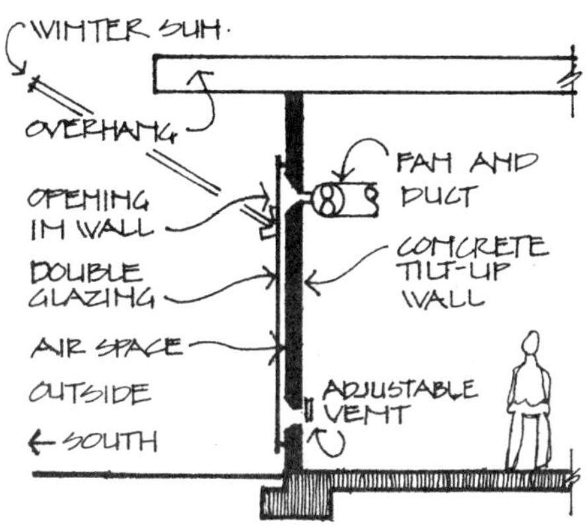

TYPE: Massive Wall with Active Backup
DESIGN: Pacific Sun and Habitec Architecture
LOCATION: Santa Clara, California LATITUDE: 37°N ALTITUDE: 50 feet
CLIMATE DATA: Heating Degree Days (65°) 2566
Cooling Degree Days (65°): 420
SOLAR RADIATION: Percent Possible Sunshine n.a.
ANNUAL DIRECT + DIFFUSE 90°S: 341,964 BTU/ft^2
ANNUAL DIRECT + DIFFUSE 60°S: 551,264 BTU/ft^2
DECEMBER DIRECT + DIFFUSE 90°S: 30,268 BTU/ft^2
DECEMBER DIRECT + DIFFUSE 60°S: 34,776 BTU/ft^2
SYSTEM DATA: Area South Glass 3000 ft^2
BUILDING HEAT LOSS: n.a.

Summer Design Temp. 97°F
Winter Design Temp: 30°F
Mean Temp. January: 49°F
Mean Temp. July: 63°F
Yearly Mean Temp: 59°F
Thermal Storage: 187,500 lbs concrete
Water Equiv. Storage/sq. ft. glass: 20 lbs/sq. ft.

courtesy Dave Bagshaw

Passive Solar Goes Underground In A Big Way
University of Minnesota 1976

Description:

The University of Minnesota Bookstore/Admissions Office illustrates what can be done even on a difficult site in a harsh climate. It includes 83,000 ft² of space which is used for the admissions and records offices and bookstore. These uses were zoned to provide sunlight and view space to those who would be working there.

The underground building is particularly attractive in Minneapolis because of the very harsh climate, with summer highs of over 100°F and winter lows as cold as - 35°F. Burying the building replaces this harsh environment with an almost steady temperature of 50°F. The energy requirement for heating due to conductive loss on a cold winter day is considerably less than half that of an above ground building with good insulation. In addition, the infiltration is greatly reduced and make up air can be very carefully controlled and heated with the exhaust air in heat exchangers.

The building makes good use of the sun for heating considering the difficult nature of the site. Glass is used for much of the south facing wall. It is tilted at 45° to improve penetration to the lower floors. Even during a winter day the heating is fully met by the sun, people, and internal loads. This is one of the factors that makes a passive commercial building so easy to do. Wise use of internal loads and only limited sun can provide full heating even in Minneapolis! A 660 ft² active collector is used to heat 8000 gallons of water in a storage tank for use at night to keep the building warm after the people leave.

The challenge in most commercial buildings will be to keep them cool. An underground structure such as this makes it much easier. The ground temperature of 50°F can be used for conductive cooling. The south window walls include louvers which block out almost all of the summer sun. Deciduous planters are also used to improve summer cooling.

In addition, the lighting system was very carefully designed to be energy efficient. This reduces the worst internal heat gain. A system of buss bars with outlets is set above the suspended ceiling. These feed power to fluourescent lights which are interchangeable with acoustic tiles. This allows very efficient lighting because lights can be moved without calling in the electricians.

Necessary additional cooling will be provided by an absorption cycle unit driven by the active solar collectors with a steam boiler in reserve. The energy conserving design and passive features allow the backup active system (only 8% of floor area) to meet 33-100% of the cooling and heating load. The savings from the energy use of the building are expected to run over $20,000 per year. Maintenance costs are also expected to be lower.

For more information contact:
Ray Sterling
Underground Space Center
Univ. or Minnesota
Minneapolis, Minn. 55455

DESIGN: Myers and Bennet Architects, Thomas Bligh, Engineer
LOCATION: Minneapolis, Minnesota
LATITUDE: 44°N
ALTITUDE: 822 ft.
CLIMATE DATA: Heating Degree Days (65°) 8382; Cooling Degree Days (65°) 500
SOLAR RADIATION: Percent Possible Sunshine: n.a.
SYSTEM DATA: Area South Glass: n.a.
Building Loss Coeff: n.a.

Summer Design Temp: 92
Winter Design Temp: -14°F
Mean Temp. January: n.a.
Mean Temp. July: n.a.
Yearly Mean Temp.: n.a.
Thermal Storage: n.a.

courtesy Ray Sterling

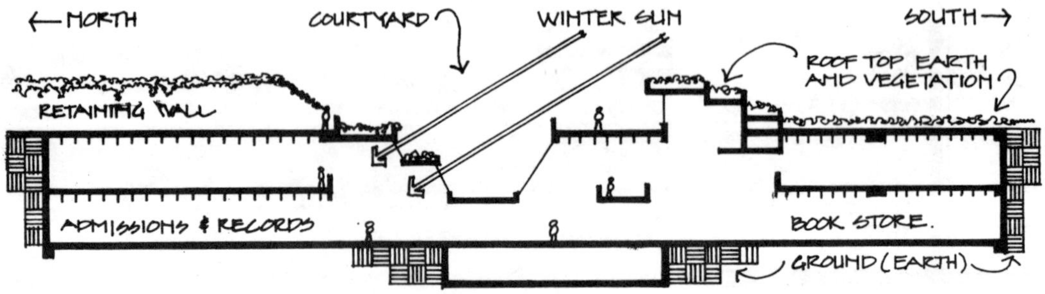

Chapter 7. Passive Solar Water Heating

1. Introduction

Although the flatplate collector water heating system is the predominant type being used today, other types of solar water systems have been developed. This chapter describes a different type of solar water heating system; the passive solar water heater. The passive solar water heater combines solar heat collection with storage. This "integrated" solar heating system consists of one or a series of tanks which are painted black to absorb heat from the sun. They are usually enclosed in an insulated box and covered with glazing material (glass or a special fiberglass) to retain the heat. The thermosiphon flatplate collector is often referred to as a passive heater but because it does not combine collection and storage it is excluded from this discussion.

The quality and cost of passive water heaters can vary considerably depending on specific design and construction details. The performance of the passive solar water heater will usually equal or exceed a flatplate collector system on a BTU/$ basis for home applications. Typically a flatplate collector system will cost from two to three times as much as a passive solar water heater. Currently, most operating passive solar water heaters are simple home-built systems constructed from inexpensive materials such as recycled electric water heater tanks. In most cases, the passive solar water heater will be the cheapest solar water heater for homeowners to install. These heaters also have good potential for mass marketing in tract houses and/or commercial applications because of their low cost.

2. History

Passive solar water heaters are not a very new idea. The first patents on this type of heater were taken out in 1891 for the "climax" heater and in 1898 for Frank Walker's system. They were used widely in California and Florida during the early 1900s. Passive solar water heaters performed well in those early days but gradually disappeared as artificially cheap natural gas became available and was aggressively promoted.

F.A. Brooks tested several passive solar water heaters in 1936 at the University of California in Berkeley. He demonstrated that tanks in insulated boxes are capable of producing hot water over 120 F. Furthermore, he found that upright tanks placed on an incline deliver hotter water than horizontal tanks. He found that passive solar water heaters were efficient solar energy absorbers but did not keep water very hot overnight.

Contemporary research has improved on Brooks' original heaters by adding nighttime insulation and reflectors to the heaters. However, recent work has often ignored Brooks' basic discoveries about tank placement and have achieved relatively modest performance. Baer, Brand and others have placed insulated lids on the box which are closed at night to reduce heat loss. Baer's lids also reflect solar radiation onto the tanks during the day. His design is known as the breadbox water heater because it looks much like the old fashioned breadbox. McCracken wraps the tanks in transmissive insulation to reduce nighttime heat losses, but this insulation also decreases the daytime collection efficiency. More recently, Marshall Hunt, Gary Starr and the Soldyne Group, Jeff Reiss and John Burton, and the author have constructed and studied breadbox water heater performance in the Davis area. These investigators have used both horizontal and inclined tanks, in a variety of designs.

Overall however, there has been very little research concerning the cost, performance, and optimum design of breadbox water heater. Although there is currently very little information describing passive water heater design, performance and cost, all the systems that have been built have worked well and it is apparent that they hold real promise for solar water heating. Passive solar water heaters will surely be the focus for increased activity in the near future with the potential for commercial kits and installations being particularly exciting.

3. Passive Solar Water Heating Systems - Basic Design Principles

There are several different types of solar breadbox water heater designs. Any of these passive solar designs will work if the following simple solar design principles are utilized:

1) The passive solar water heater must be placed in the sun, facing south to allow maximum heat gain.

2) The tank(s) should be painted black to absorb the sun's heat.

3) The tank(s) should be enclosed in an insulated box to reduce heat loss and keep the sun's heat in the water.

4) Glazing is used to trap the sun's heat by reducing convection and reradiation. (one or several layers may be used)

5) Cold water is run in the bottom and hot water is drawn off the top.

6) Optional. Insulated lids, drapes or shutters can be installed to increase the performance of the heater by preventing heat loss at night and during extremely cold, cloudy periods (usually used where freezing temperatures occur).

7) Optional. Reflectors can be added to increase the collection area of the heater. They may be combined with the insulated lids.

This chapter describes three different passive solar water heater designs: The Horizontal Tank with Lids (Baer), the Three Tank Vertical (Hunt), and the Three Tank Horizontal with Insulation (McKracken).

4. Horizontal Tank Passive Solar Water Heater With Lids (Baer)

The horizontal tank passive solar water heater with lids can be easily constructed using readily available materials and employing conventional construction techniques. Standard water tanks are painted black and enclosed in a well insulated box. Glass or another glazing material covers the south side of the box. Operational insulated lids are installed to increase performance of the heater. Incoming water from the main line enters the tanks at a low point. The solar heated warm water is drawn from a high point and routed to the existing back-up water heater.

The Soldyne Group in Davis built and tested a single tank breadbox in 1977. The plans were modified to accomodate the size of the tanks and the available materials. The box was covered with a single layer of clear polyethylene rather than a double layer of glass in order to allow easy access to the inside of the box. All the tests were conducted with this plastic glazing. Better performance would have been obtained by using a glazing such as glass or fiberglass which are opaque to infared radiation.

The single tank breadbox water heater was tested under several operating conditions but only the performance of the heater when it was connected to the domestic water heating system is presented here. The lids were opened in the morning and closed at night. Tank temperatures were monitored by thermocouples placed inside the tank. The performance of the single tank breadbox connected to the domestic water heating system is shown in the accompanying graph. The effect of drawing hot water from the breadbox is shown in several ways. First, the sharp drop in tank temperatures between 17 00 - 18 00 is due to showers taken during that period. The mixing of hot and cold water in the 30 gallon tank is most evident between 20 00 - 22 00. The tank temperature drops as warm water is drawn out and cold water at about 65 F enters. As the cold and warm water mix, the tank temperature gradually rises.

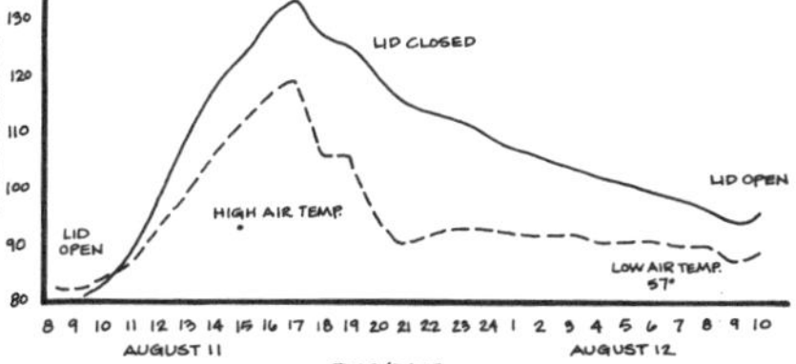

5. The Vertical Tank Passive Solar Water Heater (Hunt, Soldyne, Bainbridge)

The vertical tank passive solar water heater shown here can easily be constructed using readily available materials and conventional construction techniques. Standard glass lined water heater tanks are painted black and enclosed in a well insulated box. Glass or another glazing material covers the south side of the box. Lids or interior insulated drapes can be installed to increase performance of the breadbox. The tanks are plumbed in series with the central tank as the final step, because it will stay warmer than the end tanks.

The Bainbridge house uses a passive solar system with three 30 gallon tanks, plumbed in series and mounted at a 45 degree angle in an insulated box. The box is glazed on the south side. Insulated shutters or drapes to prevent nighttime heat loss are installed. The water is always drawn from the hottest tank with very little mixing with incoming water. The 45 degree angle mounting of the tanks serves five purposes. First, 45 degrees is close to the optimal winter collector angle for the Davis latitude. Second, it is easy to use in construction. Third, the 45 degree mounting of the tanks allows good temperature stratification to develop in the tank and insures that the hottest water is always being drawn. Fourth, mounting the tanks on an angle allows for a compact arrangement within the box. This reduces both the material cost, and the heat loss through the box and the glazing. And finally, truncating the triangular-shaped box on the top helps reduce heat loss and glazing costs and minimizes self shading. The three tanks, 16"x36" are spaced 8" apart. This spacing allows maximum exposure of the tanks during the winter.

This type of system has been in use for over 2½ years on the Hunt/Thigpen house and provides full solar hot water for 9 months of the year. More recently similar designs have been included on several other tract houses built by Ms. Thigpen. Unfortunately, no monitoring data is available yet.

The following graph presents data on several days of operation of the Bainbridge passive water heater. It is important to note that this is delivered temperature in the kitchen after an uninsulated pipe run. Further tests are planned with various insulation, drape, and shutter modifications.

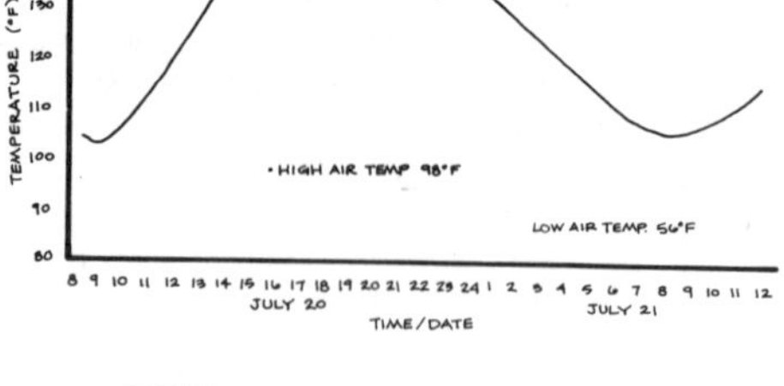

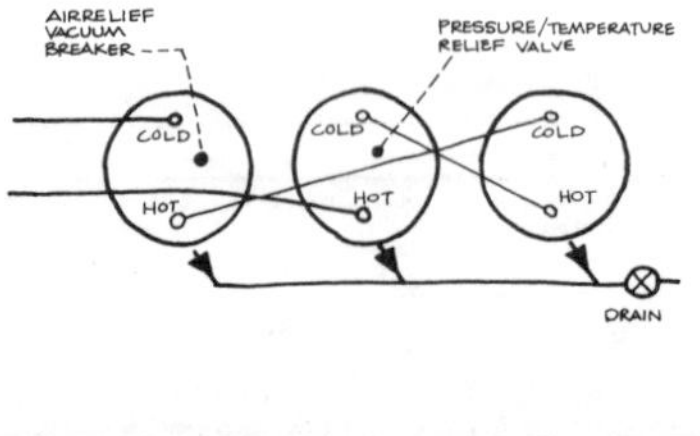

6. Horizontal Tank Passive Solar Water Heater With Insulation Wrap (McCracken)

©1978 *Horace McCracken*

The horizontal tank passive solar water heater designed by Horace McCracken is simple and straightforward to build. As described here it includes three tanks but from 2-5 have been used depending on load. The 30 gallon glass lined electric heater tanks are painted black, plumbed in series, and placed in a double glazed insulated box. A thin insulating material is used to wrap the tanks - letting the sun in but reducing heat loss at night.

Mr. McCracken has built several of these in Southern California and has collected some performance data. The accompanying graph shows system operation during early spring.

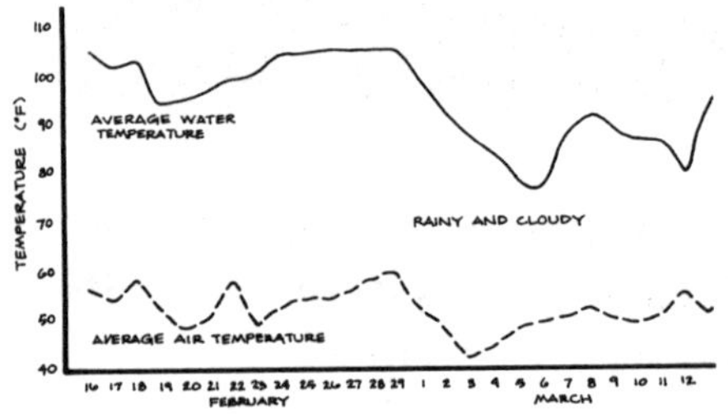

One of his heaters has been working for over three years now and provides about 80% of the hot water used by a family of three. A two tank system installed in the Imperial Valley is providing 100% of the hot water used. The following chart is Mr. McCrackens estimate of performance for Palm Springs.

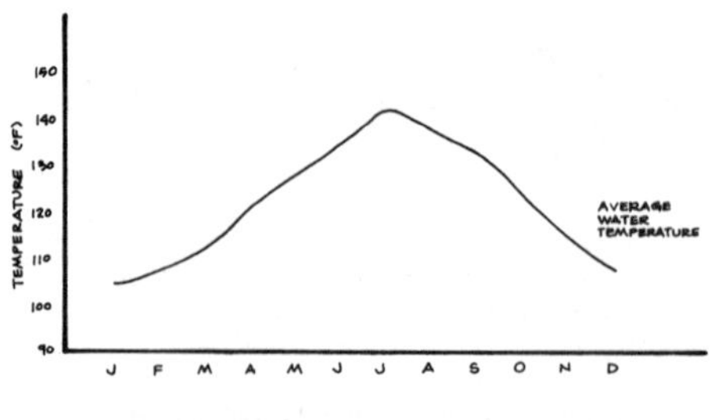

7. Placement

The passive solar water can be placed almost anywhere with good solar exposure, although it is wise to minimize the pipe run to the backup water heater.

In many houses the heater can be built on the ground, near the hot water closet. This makes installation easier and also makes shuttering more straightforward. It can also be put on a carport or patio roof if it is braced properly. If necessary it can be placed on a specially built platform.

8. Water Heater Maintenance

Excessively high hot water bills often reflect neglect of your existing hot water heater. Three areas of neglect area: a) exposing the water heater to the weather or poorly insulating the room in which it is housed, b) very high water temperature setting or faulty thermostat control, c) buildup of sediment deposits within the water heater tank. Unless a water softener exists between the cold water supply line and the domestic hot water heater, all heaters should be drained and inspected for sediment deposits annually. It makes good sense to tune-up your water heater when you install your solar water heater - or even if you don't.

9. Build Your Own Passive Solar Water Heater

The passive solar water heater is a very simple way to heat water. Water flow is provided by routing the cold water intake line through the heater and onto the backup (natural gas or electric) water heater. The backup heater should be off and bypassed in the summer. This eliminates the need for expensive pumps and/or controls. The passive solar water heater may serve as a pre-heater or provide all of the hot water needed. The following list of basic considerations can help you design and install a vertical tank passive solar water heater for your own home or business.

1. Draw plans

2. Obtain needed building permits

3. Face glazing due south, insulate box well, (6-8 inches of insulation) in an area that can support the weight of the tanks. Total system weight may be 1200 lbs so extra bracing may be required.

4. Paint interior surfaces black.

5. Mount tanks parallel to glazing rather than horizontally if space allows it.

6. Incline glazing surface and tanks at latitude angle (+10 degrees if used in winter).

7. Plumb tanks in series with the top of the last tank being the house feed. Cold water should go in the bottom with a baffle to reduce mixing. Use dielectric breaks if copper and galvinized iron are mixed. A magnesium anode will help reduce corrosion.

8. Use long thin tanks if available (mobile home tanks are good).

9. Size tanks to provide about 20-30 gallons per day per person, (typically one 30 gallon tank per person).

10. Size collector to provide at least 1 sq. ft. of glass per 2.5 gallons of tank.

11. Provide a drain valve to empty tanks if necessary. A pan under the tanks with outside drain is recomxended.

12. Provide insulated shutters or drapes and reflectors to improve performance.

13. Carefully insulate exposed pipes to prevent freezing (tanks are safe except in fairly cold areas where they should be drained in winter unless a very good shutter is used).

10. Cost

Table 1-3 itemize the costs for construction of the three different types of breadbox water heaters. Costs will change over time and may be subject to very wide local variations. The costs presented here should give you a reasonable idea of the range of costs that can be expected. It is apparent from these tables that the material cost of even new material for passive solar water heaters can be quite low. Cost can be reduced further by using recycled materials which are often free or very inexpensive. In fact, effective passive solar water heaters have been built from recycled materials (a water tank and refrigerator) for under $30. If quality is a consideration, breadbox water heaters can be built with more expensive components and materials for longer life and better performance.

Tables 1-3 show the cost for materials only because passive solar water heaters are ideal do-it-yourself projects. In this case the labor cost can be eliminated. For contractor built systems, labor cost should be about equal to the materials cost. Therefore, the total system cost would be approximately twice the cost of the materials. Cost of the materials for the hookup will vary depending on the length and complexity of the run and hookup. From 40-90 dollars is a reasonable estimate. Most state tax credits and the Federal tax credit apply to passive solar hot water heaters. The typical payback period for these heaters in the Davis area is thought to be about two years for the owner built system and four years for the contractor built system with the solar tax credit.

TABLE 1

COST ITEMIZATION
TWO TANK PASSIVE SOLAR WATER HEATER

Item	Amount	
2x4's	108 ft	34.20
Plywood - ⅜ AC exterior	5 sheets	40.80
Insulation - R-11 fiberglass	50 ft	9.50
Insulation - cellulose	2/3 bag	2.65
1x3's	84 ft.	10.00
T-hinges	3 pair	6.60
Wood Sealer	2 qts	4.95
Paint - exterior latex	3 qts	8.65
Caulk	2 tubes	2.45
Silicone seal	1 tube	5.20
Glass - 2x3 sheets	14 sheets	112.00
Plumbing		
Hose bibs	3	7.50
Couples (¾)	8	7.10
¾ plugs	5	2.45
¾ L's	4	1.90
Assorted nipples	6	3.00
Pipe (24x¾)	2	3.40
Tanks (30 gallon)	2	82.50
		$344.85

***Constructed by the Soldyne Group in Davis, California, Summer 1977.**

TABLE 2

COST ITEMIZATION
THREE TANK VERTICAL PASSIVE SOLAR WATER HEATER (1978) BAINBRIDGE

Item	Amount	
2"x4's	124 feet	43.40
Plywood ⅜" ext.	4 sheets	33.20
Insulation - R-11	1 roll	11.00
Insulation - Tech Foam 2"	1½ sheets	33.00
paint, ext.	2 qts	11.40
paint, black	1 qt	6.00
f-glass glazing	42 sp. ft.	63.00
flashing, misc.		15.00
tanks, 30 gal.	3	135.00
dial. flex connectors	10	40.50
nipples, galv. ¾	9	2.25
PT relief	1	4.59
Misc. copper fittings	15	9.00
copper pipe ¾	25'	23.00
hose bib	1	2.49
caulk	2	4.40
misc. nails, screws, etc.		25.00
		$462.23

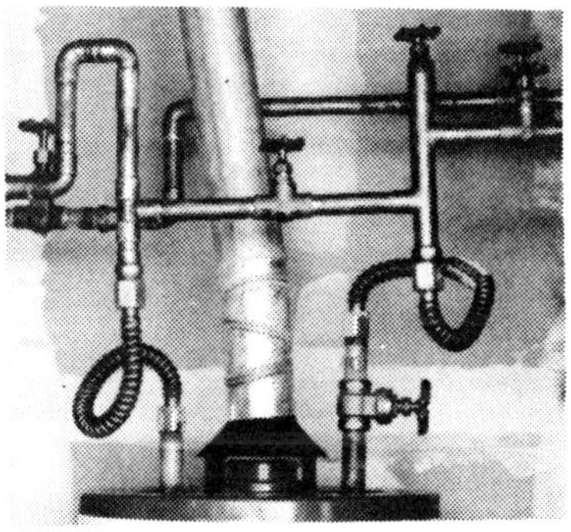

TABLE 3

COST ITEMIZATION
THREE TANK HORIZONTAL PASSIVE SOLAR WATER HEATER (1978) MCCRACKEN

Item	Amount	
2x6	50'	19.00
1x4	12'	1.20
1x3	22'	3.74
½" plywood	4 sheets	63.52
insulation	6 batts	17.00
aluminum trim & flashing	various	35.00
paint ext	1 qt	4.30
paint black	½ qt.	2.25
tanks, 30 gal.	3	231.00
¾ nipple, stainless	9	13.05
¾ el, stainless	3	19.30
flex. copper c	2	14.00
union - brass ¾th/sw	2	8.48
PT relief	1	5.65
glass, dbl. strength 24"x30"	12	88.20
silicone sealant	1	5.95
misc., (nails, tape, etc.)		25.36
		557.00

11. Conclusions

Three tank and two tank passive solar water heaters have performed substantially better than one tank breadbox water heaters in the testing that has been done. Full solar water heating is provided for 9 months in Davis (79% sun, 2418 degree days). This is comparable to thermosiphon flat plate performance. Even a small simple one tank system may prove effective as a preheater and will help reduce utility bills. A full size passive solar water heater can provide much of the annual hot water demand for a household in most climates of the U.S.

Sizing and system design criteria are still fairly crude but will be in the ballpark if collection area is based on flat plate considerations. A detailed investigation is needed to determine the relative performance of the passive solar heater and flat plate heater.

There are many variations to the designs presented here. As passive solar water heating becomes more popular, it will stimulate a flurry of research and development activity that will present many different design options. Currently, the field is wide open for experiments and innovative designers.

One possible design concept which has not been tried or tested would be more simple and considerably less costly (per unit of hot water) than the three tank residential system. It probably would include a single long tank of 90-120 gallons (either round or rectangular) built into the roof. The tank should include brackets for hanging between rafters and braced off a wall. This type of system might be retrofitted under an eave overhang. The upper part of the tank would be insulated to contain the heat in the water and reduce nighttime heat loss. This type of system could be developed for small scale (residential) use and larger scale (multi-family, commercial and industrial) applications.

A great untapped market exists for commercial passive solar water heater kits and complete water heaters.

Further Reading: Passive Solar Water Heaters

Akers, D. (1978) "$30 Homestead Solar Water Heater", **Mother Earth News**, May-June 1978 No. 51.

Baer, S. (1975) **Breadbox Plans,** Zomeworks, Box 712, Alburquerque, New Mexico 97103, $5.

Bainbridge, D. and Reiss, J. (1978) "Breadbox Designs," **Alternative Sources of Energy**, #34.

Brooks, F.A. (1936) **Solar Energy and Its Use for Heating Water in California**, UC Berkeley Agric. Exp. Station, Bulletin 602. Available from OAT, 1400 Tenth Street, Sacramento, CA 95814.

Butti, K. and Perlin, J. (1977) "Solar Water Heaters in California 1891-1930," **CoEvolution Quarterly**, Fall.

L.A. Times Home Magazine, May 1, 1977.

Lillywhite, M. ed. (1978) "Domestic Hot Water Systems Section," **VII 2nd Nat'l Passive Conference Proceedings**.

Lucas, T. (1976) **Economical Solar Water Preheater**, 10371 Stone River Court, Fountain Valley, CA 92708.

Maeda, B. (1978) Personal communication, r/e simulation runs of breadbox performance, Davis Alternative Technology Associates, P.O. Box 503, Davis, CA 956)6.

McCracken H. (1976) **Passive Solar Water Heater,** 329 W. Carlos, Alturas, CA 96101, $6 ppd.

Missouri Department of Natural Resources, (1978) **Solar Water Preheater.**

Morgan, E. (1976) "Solar Bubble," **Popular Science**, September, P. 101.

Staff (1978) "Analysis of Energy Conservation Techniques: Davis Retrofit Program," City of Davis/Sedway-Cooke.

Starr, G., Melzer, B. et al., (1978) **Resource Conservation for Residential Buildings**, N.S.F. Grant 77-05558, Davis, CA.

Union Electric (1978) **A Simple Approach to Solar Heating of Domestic Hot Water**, Union Electric Co., St. Louis, MO.

Von Vleet, A. (1977) "Heating Water In A Breadbox," **Alternative Sources of Energy**, (25):21-25, April.

Zweig, P. (1978) **Inexpensive Do-It-Yourself Solar Water Heaters or Preheaters,** PCRI, 3875 Jackson Street, Riverside, CA.

Chapter 8: Daylighting

"Daylight is a gift of nature. As civilized man learns to use artificial light sources which free him from total dependence on daylight, he also learns to appreciate the value of daylight and becomes aware of its special advantages. The idea that man evolved by adapting to his natural environment is a basic feature of 20th century thought. This concept implies that human vision is adapted to daylight, and that consequently artificial light necessarily involved a certain degree of unfavorable adaptation."

R. G. Hopkinson *et al*
Daylighting (1966)
Heineman

1. Introduction

In 1972, more than 20 percent of the electricity generated in the United States was used for lighting, FEA (1975). If this energy had been wisely spent to provide comfortable and safe working conditions this very high use might be justified; in fact, however, much of this energy is spent for over-lighting that in many cases causes discomfort and increased hazard. In addition, this over-lighting greatly increased the use of energy during the summer for air conditioning to offset the heat added by the lights. Proper lighting design with full use of daylighting could save substantial quantities of energy and money while providing more comfortable and healthful conditions for workers throughout the United States.

2. Seeing and Lighting

The basic goal of lighting should be the provision of a comfortable, pleasant, and safe visual environment for man. This is not as simple as one might expect because the human eye and brain are very complex and adaptable. These make good lighting design both difficult and easy at the same time.

There are several concepts that must be understood in the discussion of lighting. The first is illumination, or the amount of light spread over a surface. This may be described in the English notation by the term foot candle or in the metric system as a lux or lumen/sq. m. [1 fc = 1 lm/sq.ft. = 10.8 lux]. Much of the discussion that follows uses fc as the sole reference; however, it is only the most convenient of several factors that must be considered in good lighting.

Brightness is a factor of the level of illumination and the reflectance of the surface we are viewing. A white surface may have a reflectance of nearly 100 percent while a dark black surface may reflect only two percent. Brightness is the product of illumination multiplied by the reflectance and is also known as "luminance." The common measurement is the foot-lambert.

Perceived brightness may vary considerably from the physically measured brightness. For example, a car headlight is apparently very bright at night yet very dull during the day even though the measured brightness is the same. Color may have a similar effect, Hopkinson (1969).

Thus, in discussing brightness we must discuss it subjectively - for what we see. The human eye can see from around .0001 to 10,000 foot candles. It is very adaptable and can be comfortable over a very broad range of lighting intensities. Research has repeatedly shown that low light levels will not cause eye damage. Variety is essential for visual comfort and must include movement, intensity, and color. The environment in which we evolved provided this stimuli and we require it for comfort and good visual performance.

Natural lighting is desirable because using the sun matches the spectrum of light our eyes evolved with exactly. The spectrum of light from incandescent lights, fluorescent lights, mercury vapor, and sodium vapor lights are considerably different and appear more and more to cause adverse psychological and/or physiological responses, Dubos (1964), Ott (1976), Clark (1974). Unfortunately, these impacts have received very little study and remain largely speculative, Fitch (1972).

3. Lighting Standards and Research

"It is unfortunate that American engineers do not appreciate the value of experimental checks on their computations. A system of computations might be internally consistent and still be in conflict with experience on the common sense level. In such a case it is of theoretical value only, which lighting is not.

The British lighting code which, like the codes of other European countries, is more conservative than the American one, has apparently taken into account much more experimental data than has the United States' code. As a result, the human being figures more strongly in the British approach and is not completely eliminated in favor of the visual task as a criterion." Leslie Larson.

Lighting standards have risen drastically in the United States largely **in spite of** the research that has been done on human comfort. The Illuminating Engineers Society has been at the forefront of the revisions, largely in response to strong pressure from the manufacturers of lighting equipment, Clark (1974). The history of the N. Y. School lighting standards illustrates the magnitude of the changes that have occurred. Before 1910 the standard was 3 fc, between 1910-1950 it jumped to 30 fc, and it has risen since then to 150 fc, Steadman (1975).

As one might expect, this is a uniquely American phenomenon. European engineers have continued to rely on the concept of human comfort rather than the idea of maximum potential performance reading very poor material. The biggest rise in standards was based on a very narrow interpretation of research by Dr. R. H. Blackwell. His research in the 1950's led to a recommendation for light levels for various purposes based on 99 percent accuracy of visual task completion. He reported that foot candle levels of only 0.60 fc were sufficient for reading a clean copy of 12 pt type (the quotes in this paper). For equal accuracy reading an original typed with a very poor ribbon almost 3,000 fc were required. The I. E. S. focused on the higher levels of illumination, even though subsequent research by Dr. M. A. Tinker found that visual performance in reading newsprint did not increase after 7 fc. Other studies have demonstrated that readers prefer much lower levels than are now prescribed, Larson (1964).

The European Standards are based more clearly on empirical studies of comfort and standards have used only 90 percent rather than 99 percent accuracy as a criterion. Their stated goal has been "lighting which would permit efficient work in a pleasant environment." A comparison of the standards in different countries is very instructive, Table 1. There seems little question that the minimum levels recommended by the I. E. S. in the U.S. are unjustified.

The fact that the recommended standards in the U.S. vary so widely from the rest, and that they have risen most markedly, suggests that they needn't be taken very seriously. The wide variation in standards also reveals the highly subjective rather than scientific nature of the standards and the standards setter.

A revised lighting standard prepared by the FEA is included below, Table 2, FEA (1975). Note that it is for maximum lighting levels. In fact, it is still very high for most purposes.

TABLE 2: FEA'S RECOMMENDED MAXIMUM LIGHTING LEVELS

Hallways or corridors	10 +or- 5 fc
Work and circulation areas surrounding work stations	30 +or- 5 fc
Normal office work, such as reading and writing (on task only), store shelves, and general display areas	50 +or- 10 fc
Prolonged office work which is somewhat difficult visually (on task only)	75 +or- 15 fc
Prolonged office work which is visually difficult and critical in nature (on task only)	100 +or- 20 fc

I have prepared a revised standard based on a review of the literature. This is included below in Table 3.

TABLE 3: RECOMMENDED LIGHTING LEVELS

Hallways or corridors	3 fc
Work and circulation areas surrounding work stations	10 fc
Normal office work (with task light available)	30 fc
Prolonged office work of detailed nature (with task light available)	50 fc

It is important to maintain a ratio of 10:3:1 for work, foreground, and background lighting levels. If background lighting levels are high, discomfort and reduced performance result because this ratio cannot be maintained.

Table 1: RECOMMENDED LIGHTING STANDARDS from Larson (1964)	**U.S.A.**	**SWITZERLAND**	**FINLAND**	**BELGIUM**	**SWEDEN**	**FRANCE**	**GT. BRITAIN**	**AUSTRALIA**
Most Difficult Seeing Tasks Finest precision work. Involving: finest detail; poor contrasts; long periods of time. Such as: extra-fine assembly; precision grading; extra-fine finishing.	**926- 1852**	**over 93**	**93- 185**		**93- 185**	**139- 278**	**139- 278**	**over 185**
Very Difficult Seeing Tasks Precision work. Involving: fine detail; fair contrasts; long periods of time. Such as: fine assembly; high-speed work; fine finishing.	**463 926**	**27- 93**	**46**	**46- 94**	**27- 46**	**65- 139**	**65- 139**	**65- 139**
Difficult and Critical Seeing Tasks Prolonged work. Involving: fine detail; moderate contrasts; long periods of time. Such as: ordinary benchwork and assembly; machine shop work; finishing of medium-to-fine parts; office work.	**463 92**	**14 28**	**28**	**23 46**	**28**	**28 65**	**28 65**	**28 65**
Ordinary Seeing Tasks Involving: moderately fine detail; normal contrasts; intermittent periods of time. Such as: automatic machine operation; rough grading; garage work areas; switchboards; continuous processes; conferences and life rooms; packing and shipping.	**46- 93**		**14**	**9- 23**	**14**	**14- 28**	**14- 28**	**14- 28**
Casual Seeing Tasks Such as: stairways, reception rooms; washrooms, and other service areas; inactive storage.	**18- 28**	**4-8**	**8**	**5-8**	**4-8**	**7-14**	**7-14**	**7-14**
Rough Seeing Tasks Such as: hallways; corridors, passageways; inactive storage.	**9-18**		**4**	**2-3**	**2**	**3-6**	**3-6**	**5-7**

4. Daylighting

"The typical illumination specialist has a background in electrical engineering: he is thus all too prone to rely upon artificial light sources exclusively, paying little or no attention to daylighting."
J. M. Fitch (1972).

Daylighting was an integral consideration in building design and influenced building orientation and fenestration very strongly until recently. It is unfortunate that it has now dropped from consideration as it provides benefits other than the inestimable psychological values of daylight. Foremost is the reduction in energy use for lighting. This leads to direct dollar savings that will grow as energy prices spiral upward. In addition, natural lighting can reduce heat gain from lighting to 1/4 that obtained with fluorescent lights and as little as 1/6 that from incandescent lights. This can lead to a substantial reduction in air conditioning load and consequent energy and financial savings.

The ultimate source of all daylight is the sun, yet only a part of the light use for natural lighting is received as direct sunlight. The rest arrives as indirect or diffuse radiation reflected off dust or water particles in the atmosphere or reflected radiation bounced off other structures or features of the landscape. The proportion of each varies as a function of both building orientation and design, variation in ground surface, and also varies as the sun's path and atmospheric clarity change.

Using natural light is not difficult, yet it has been widely ignored in the United States because of cheap electricity and a fascination with artificial lighting. Great Britain and Europe have continued to use daylight and an extensive literature is available for use in daylight design, BRS (1966), Hopkinson and Kay (1972), Walsh (1971), Hopkinson et al (1966).

Models are particularly useful in lighting design work. A scale model of a building is a true photometric analogue and is much easier and more accurate to use than any sort of calculation, Hopkinson (1958), Models of 1/20 to 1/4 scale are relatively inexpensive and easy to construct yet are large enough to provide very useful information.

5. The Techniques of Daylighting

The actual design of a lighting system must meet the many objectives listed below, Table 4. The methods for meeting these objectives are varied and allow great flexibility for a sensitive and skilled designer.

TABLE 4: LIGHTING OBJECTIVES

cisual comfort and efficiency for all users

maximum use of natural lighting

integration with other building systems

minimal lifecycle energy use and cost

low initial cost

maximum flexibility

minimal undesired heat gain

minimal use of non-renewable energy sources

automatic control of switching to reflect natural light levels

minimal glare

A. WINDOWS

Windows offer one of the better methods for providing natural lighting because they are almost always incorporated in the building design. If used properly they can provide "free" natural light. Orientation, placement, and control are essential for good natural lighting—without glare, excessive heat gain, or great variation in brightness. Proper design can also ensure that the windows also provide needed cross ventilation for natural cooling—an added bonus.

Windows should have narrow glazing bars and frames and window surrounds should be angled when possible. Diffusers can be used to reduce glare where needed. Windows should be shaded during the summer in hot areas. In very hot areas, translucent insulated shutters or shutters with light ports may be desirable to reduce conductive heat gain.

Effective placement of windows will yield the best light for a given window area. Thin tall windows placed near walls wash the walls with light and provide pleasant lighting without excessive contrast, Figure 1.

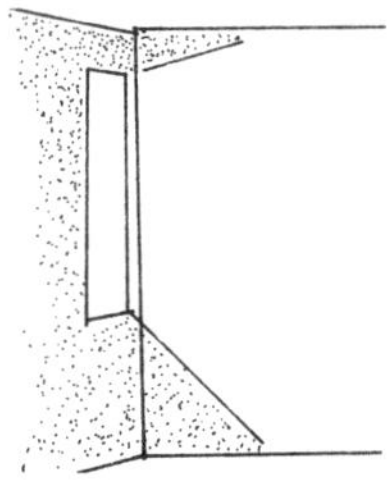

FIGURE 1: LIGHTING WITH TALL CORNER WINDOWS IS VERY EFFECTIVE

High windows admit more light and are also desirable for venting hot areas. Yet disability glare will occur unless they are isolated with some form of screen. One of the best methods of screening is the use of a light shelf, Figure 2.

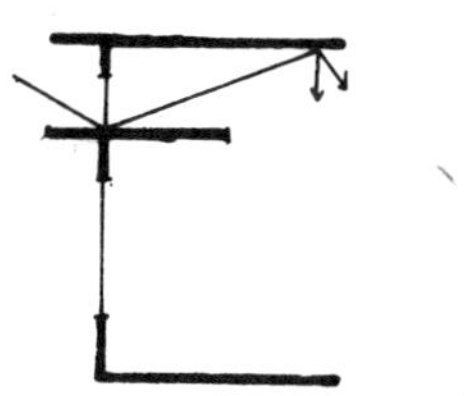

FIGURE 2: THE LIGHT SHELF IS A PROVEN AID IN NATURAL LIGHTING

The light shelf also helps spread light onto the ceiling and increases light penetration into the room. With a light shelf, adequate natural lighting can reach up to 24 feet from the window, Hopkinson and Kay (1972). The initial cost of window and light shelves will be lower than for artificial lighting if the cost of added generating capacity and transmission losses are included in cost of artificial lighting.

B. SKYLIGHTS

"A common mistake in comparing the economy of daylight utilization with that of electric lighting is made by many air conditioning people.

The normal procedure is to assume a fixed electric lighting load and consider any daylighting as an additional heat load. This assumption is erroneous and does not recognize the advantages of daylight utilization."

J. W. Griffith

Skylights are in many cases the best method of natural lighting. They should face north or should be shaded in the summer to reduce heat gain using the concept of the north facing shade developed by L. W. Neubauer and refined by Living Systems, Hammond *et al* (1976). In addition, the skylight should be double glazed, preferably with a diffusing inner surface, and should have the surroundings painted bright white, Figure 3.

FIGURE 3: SKYLIGHTS ARE VERY EFFECTIVE FOR NATURAL LIGHTING

The following chart makes a comparison of various features of two types of lighting, skylight and fluorescent lighting, Table 5, Evans (1962). Note that the cost of additional heating/cooling must be added to both systems of lighting. (Note also that the data from the chart are from 1962.) All of the costs, and especially the cost of power, will continue to increase rapidly. Skylights still have **no** power cost. To be energy efficient, skylights must be carefully designed for lighting. Therefore, skylights intended for view or purely architectural reasons may not pay for themselves.

Although skylights are almost 100 percent efficient as the light enters, eventually that light does convert to heat and cooling may be required to offset the heat gain. Likewise, infiltration and conduction of air through skylights may call for additional heating or cooling. However, electric light has only a 20-50 percent initial efficiency; the other 50-80 percent is given off directly as heat gain that must be met with additional cooling. Electrical lighting, even with fluorescent lights, requires much more cooling than properly used skylights will.

This efficiency is reflected in reduced costs of lighting using skylights rather than artificial lights, Table 6. This is particularly important in this era of rising energy prices. A life cycle comparison of lighting cost makes natural lighting look even better.

TABLE 6: COMPARATIVE COSTS OF SKYLIGHTS AND FLOURESCENT LIGHTS

Cost Consideration	Source of Illumination	
	Skylights[1]	Fluorescent Light Units[2]
First cost of lighting installation (less lamps)	$480.00	$657.00
Uniform annual cost of recovering first cost at 3% for an operating period of 25 years	27.50	37.75
Annual cost for insurance at 1.2% of first cost	5.76	7.89
Annual cost for lamps	- - -	8.72
Annual labor costs for cleaning and relamping	3.00	10.80
Annual power cost	- - -	59.24
Total annual lighting cost	$ 36.72	$124.40
Annual cost per foot-candle	.74	2.43

[1]Six skylights, 38x36 in., 43% transmission with louver shade; average lighting level: 49 foot-candles.

[2]Eighteen fluorescent units, two 40-watt lamps, 45 degree louvers; average lighting level; 51 foot-candles.

Efficiency can be increased by using curbs with U-factors as good or better than ASHRAE 90 for roof U-factors. Curbs should be well-insulated with little or no infiltration. Double domes also cut down conduction and infiltration. A study by the Center for Industrial and Institutional Development tested double dome skylights with no infiltration and insulation to ASHRAE 90 for roof U-factor in various locations in the United States with the results shown in Boyle (1975).

A final concern, especially from the architectural point of view, is control of the brightness of skylights. Incoming light needs to be diffused over as large an area as possible. This problem can be handled in the choice of a more or less translucent material for the skylight and with a special diffusor lens. According to Boyle (1975), ". . . skylight energy performance is even better in those applications using a lens and all clear dome sheeting." Such problems are readily worked out by experimentation and careful design.

Skylights can and should be designed to provide needed solar heat gain in the winter for residential and small commercial projects. This is perhaps best accomplished with a reflector which serves to increase heat gain in the winter, yet provides shade in the summer. A once a year adjustment in angle or tilt may be all that is required, Figure 4.

FIGURE 4: AN ADJUSTABLE SKYLIGHT SHADE/REFLECTOR MAY BE DESIRABLE

Even better performance can be achieved if more sophisticated skylight systems are used. A tracking mirror can be used to concentrate and reflect added sunlight through the skylight. Use of special infrared screens and filters can ensure that the skylight provides almost cold sunlight for maximum efficiency, Figure 5, Parker (1978).

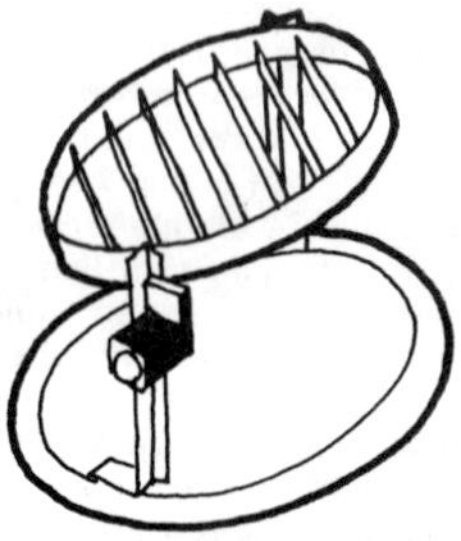

FIGURE 5: THE COLD SUN SKYLIGHT

TABLE 5: EFFICIENCY COMPARISON OF SKYLIGHTS AND ELECTRIC LIGHTING

Source	Light Output		Air Conditioning Load
	Lumens/Sq. Ft.	Lumens/Watt	Tons/100,000 Lumens
Daylight through acrylic plastic, high transmission	5,270		.27
Daylight through acrylic plastic, medium transmission	3,110		.27
Fluorescent light	- - -	60	.63
Incandescent light	- - -	20	1.90

Note: .033 more tons of air conditioning per 100,000 lumens of light at an outside temperature of 100 degrees and an inside temperature of 78 degrees should be added to the skylight air conditioning loads.

From Evans (1962)

"Since one of these fundamental needs is for access to daylight and to the sun, the architect, when considering the general form and character of a building will have to bear in mind this requirement for daylight..."

R. G. Hopkinson

6. An Integrated Lighting System for New Construction

Yet, although day lighting can meet much of the full lighting needs in many areas, artificial lighting must often be installed to supplement the day lighting in the winter. In Sacramento, for example, natural lighting will meet only about 75 percent of the annual lighting needs for a building used between 8 a.m.-5 p.m., Hammond et al (1976). Thus a back-up system is needed, and careful integration of design becomes crucial.

The design of supplementary artificial lighting must also meet the objectives identified earlier. Maximum comfort, flexibility, and minimal heat gain and cost are again the guiding criteria. As energy prices continue to rise, the cost of artificial lighting will cause increasing concern and will justify the extra time devoted to development of better lighting systems, Stein (1973).

The artificial lighting must meet needs for both area and task lighting when natural lighting is not sufficient. Fluorescent lights are more energy efficient, hence more economical and cooler, than incandescent lights and are desirable for that reason. Full daylight spectrum fluorescents are available and should probably be used whenever people will be exposed consistently, OH (1976). Background lights should be photo cells switched to turn off when natural lighting levels are adequate. Savings of 50 percent or more are possible with automatic switching, TAL (1978).

Ideally, these fluorescents will provide primarily reflected rather than direct light (with accompanying glare). A possible fixture is shown below, Figure 6, as developed by Living Systems for OB 1, Hammond (1976).

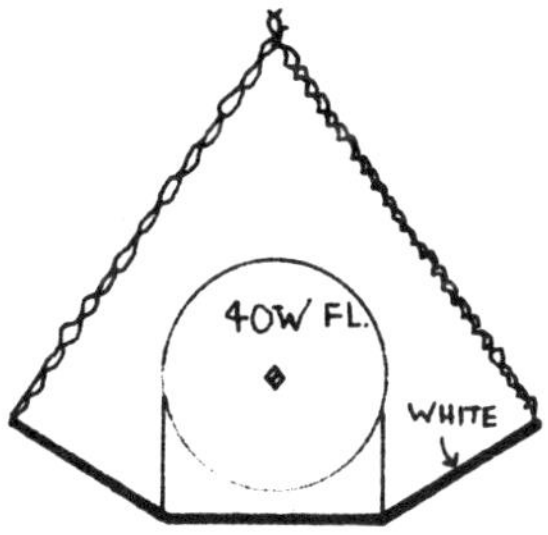

FIGURE 6: A LOW COST, LOW ENERGY USE LIGHT FIXTURE

With 40 watt bulbs suspended on a standard grid with proper ceiling and wall reflectors, 15 fc can be provided at only 0.75 watts/sq. ft. This provides a savings of 2-3 watts/sq. ft. over standard practice.

Incandescent lights are more flexible and cheaper and are well-suited for task lights. Although there has been a lack of suitable lighting systems for task lighting the energy crisis has brought renewed interest and some tolerable designs are now available.

The design of interior partitions and interior decoration must realize and accept repsonsibility for their potential impact on lighting. High reflectances are essential for the ceiling and upper walls and desirable for partitions, lower walls, floors, and furniture. However, desk tops should be chosen for moderate reflectance to reduce potential glare problems, (which should be minimal because of proper lighting design).

In addition, a "Lighting Manual" should be prepared for each building built to ensure that task and background lighting is properly set when rearrangements take place. This will help minimize energy use while maximizing human comfort and that is about the best we can ever hope to do.

7. Retrofitting Existing Structures

Improving lighting systems in existing structures is possible and can offer equal benefits, Table 10. One of the first programs in any retrofit is a light survey and analysis of energy use. Light levels can be measured using a foot candle meter ($160) or more approximately with a lightmeter set at ASA-100, EV=5 at 10 fc, EV=7 at 30 fc, EV=9 at 50 fc, EV=10.5-11=70 fc and EV=12 at 200 fc. The lightmeter readings are only approximate because the lightmeter is set to measure the spectrum that film responds to - rather than the human eye. This provides a clear identification of the problems and opportunities of the structure and sets some boundaries for the economic trade-off of remodeling.

In recent buildings, with electrical use of 4-5 watts/sq. ft. immediate improvement in comfort can often be achieved by removing one half or more of existing bulbs or lowering wattage an equivalent amount. In some cases, dimmers or variac transformers can be installed to control existing circuits. This can provide very long life of bulbs and may in cases be cheaper to install than new light bulbs.

A typical office now includes sixteen 40 watt flourescent bulbs, providing 150-200 fc. If these are replaced with four 35 watt-mizers (TM) to yield 30-50 fc, the building operation will save: $46.96 in 1975, $93.92 in 1982.

The next steps begin to cost more money and may not be cost effective in the short run at this time. However, when life cycle costs are considered, most of these steps will be very cost effective. Circuits can be rewired to provide user control, task lighting can be provided, and skylights or new windows may be installed or modified.

One of the most common and most difficult cases involves remodeling an office with a luminous ceiling. Here little may be possible except revised wattage and switching to more efficient bulbs and provisions of task lights to provide the desired ratio of light 10:3:1.

Renewing older buildings can be easier because windows are often placed better. Here switching to lighter decorations and wall and floor finishes and improvement of the windows themselves may provide reasonably good daylighting. The addition of a limited number of skylights and windows should allow much better use of daylight and greatly reduced use of artificial lighting, Lynes (1968).

8. A Challenge

The use of daylighting offers enormous potential for saving energy and money while increasing comfort and satisfaction with the visual environment, Steadman (1975). Yet for various reasons, primarily a result of breakdowns in education and professional training, the science of daylighting is in its infancy in the U.S. Therefore, if you desire to use daylighting you may have to do some of the design work yourself - or search long and diligently for a competent daylight specialist. The savings and benefits are so great that you will be very glad that you persevered. The references, appendix, and list of consultants in Chapter 11 should help get you started.

Building Research Station (1966), **Integrated Daylight and Artificial Light in Buildings**, Digest #76, Second Series, H.

Clark, W. (1974), **Energy for Survival**, Anchor, N.Y.

Dubos, R. (1964), **Man Adapting**, Yale Press, New Haven.

Evans, B.H. (1962), **Solar Effects on Building Design** Pub. #1007, Building Research Institute.

Federal Energy Administration (1975), **Lighting and Thermal Operations: Guidelines**, USGPA, Washington, D.C. 20461.

Hammond, J., Hunt, M., Bainbridge, D., and Plumb, J. (1976), "New State Office Building, Eighth and P: Design Considerations," Living Systems for the Office of the State Architect.

Fitch, J.M. (1972), **American Building: The Environmental Forces That Shaped It**. Houghton-Mifflin.

Griffith, J.W. (1961), Skylights(?) Southern Methodist Univ.

Hopkinson, R.G. (1958), "Model and Measurement in Lighting Research," **Discovery 19**, p. 232.

Hopkinson, R.G., Petherbridge, P., and Longmore, J. (1966), **Daylighting**, Heinman.

Hopkinson, R.G. and Kay, J.D. (1972), **The Lighting of Buildings**, Fabes, London.

Hopkinson, R.G. (1969), **Lighting and Seeing**, Heinman.

Larson, L. (1964), **Lighting**, Whitney.

Lynes, J. (1968), **Principle of Natural Lighting**, Elsevier, N.Y.

Ott, J. (1976), **Health and Light**, Pocket Books.

Parker, J. (1978) "The 3rd Generation Cold Solar Light is Sealed Capsule Unit" **Alternative Source of Energy** #33, August.

Steadman, P. (1975), **Energy, Environment, and Building**, Cambridge University Press.

Stein, R. (1973), "A-E Team Can Short Circuit Major Crisis," **Consulting Engineer**, March.

Technical Applications Lab (1978) "Photocell switching savings to 94%" reference from Bob-Harman, P.P.G.

Walsh, J.T. (1961), **The Science of Daylight**, MacDonald.

A Simplified Process for Daylight from Haggard et al (1978).

Some designers prefer to develop an overall theme for integrating natural light into their project. Yet others would rather base the overall scheme more on other design factors and then approach each individual space as a natural lighting design problem. At either scale, it is assumed that the typical designer will have by this time:

* a general idea of image qualities of interior/exterior environs
* the general compass orientation of the building
* awareness of surrounding landscape and buildings
* a good idea of activities within each building area

Start the following process during schematic design. It may be recycled through many times as design trade-offs become more apparent. Each step will ask you to draw or compare - do it at the level you are designing - overall or space specific. Sketches of a typical situation provide general guidance below.

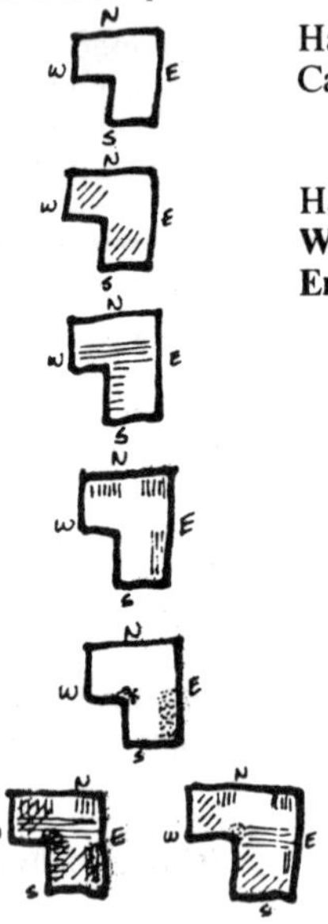

1. Draw the general plan view of the area under consideration. Label compass directions.
2. Where in the plan do you desire natural light for stationary tasks, such as reading? Draw in the region with a 45 degree cross-hatch.
3. Where in the plan do you desire natural light to illuminate circulation routes? Draw the region with a horizontal cross-hatch on an unmarked general plan. Don't redraw step 2.
4. Where in the plan do you desire general background lighting? Draw the region with a vertical cross-hatch on another unmarked plan. Again, don't redraw earlier steps.
5. Where in the plan do you desire visual accent lighting? Draw this region with a dot pattern.
6. Now draw a composite sketch showing all lighting goals.

After you have done this, look carefully at what you have. How realistic is it? Cut down all "excess" light requirements as much as possible. Remember, task lighting and lighting for circulation routes are brighter than general background or accent lighting. Therefore, light for those first requirements might be utilized for less stringent needs. After you have re-examined your lighting goals, draw a new composite diagram reflecting the changes. If you were working at the overall scheme level, later as you start dividing up the space you can meet more specific criteria by repeating this process. If you were already at a small space level, now you can repeat the process for some other areas in your project.

The following examples show the light effects of windows ranging from 1-9 (9=brightest). The following performance is desirable:

9-7 meets task lighting performance
6-5 meets movement lighting performance
4-3 meets background lighting performance
2-1 meets visual accent lighting performance

The values are indicated only for daylight, stated at ceiling height, table height, and floor level. Sunlight is not computed here, although it can become a very useful bonus (refer to sunlight chart). Relative spatial proportions count, not absolute dimensions, so the value intervals are indicated on a relative scale. By combining many of the configurations, while utilizing the rules of thumb for diffusion, glare, and sunlight, you are well on your way to successful schematic design.

Authors note:

The value of a model cannot be overstated. A tilting, rotating base and use of a remote sensing photometer can simulate performance throughout the year.

Further Reading

Beckett, H.E. (1974) **Windows**, Van Nostrand Reinhold Co., New York.

Collins, Belinda L. (1975) **Windows and People: A Literature Survey**, National Bureau of Standards, Washington, D.C.

Dorsey, Robert T. (1972) "On Uniting Esthetic and Engineering Approaches to Lighting," **Lighting Design and Application**, September, pp. 23-27.

Evans, B. **Daylighting and Architecture**, College of Architecture and Urban Studies, Virginia Polytechnic Institute and State University, Blacksburg VA 24060.

Haggard, K. et al (1978), **Passive Solar Design**, Cal Poly San Luis Obispo

Hastings, S.R. and Crenshaw, R.W. (1977) **Window Design Strategies to Conserve Energy**, NBS, BSS104, USGPO.

Olgyay, Aladar & Victor Olgyay. (1976) **Solar Control and Shading Devices**. Princeton University Press, Princeton, N.J.

Phillips, Derek. (1964) **Lighting in Architectural Design**. Mcgraw-Hill, New Yor,

Pohl, Jens. (1974) **The Layman's Lumen**. Cal. Poly. State U., San Luis Obispo.

Rosenfeld, Arthur H. & Stephen E. Selkowitz. (1976) "Beam Daylighting: Direct Use of Solar Energy for Interior Lighting." **I.E.S. Conference Proceedings**.

Tools

Building Research Station: **Daylighting Calculator** (perhaps most useful in cloudy areas similar to England).

Libby, Owens, Ford: **Sun Angle Calculator**, Daylight overlays and descriptive book, "How to Predict Interior Daylight Illumination" are well done, from L.O.F. 811 Madison Ave, Toledo, Ohio 43695.

Windows

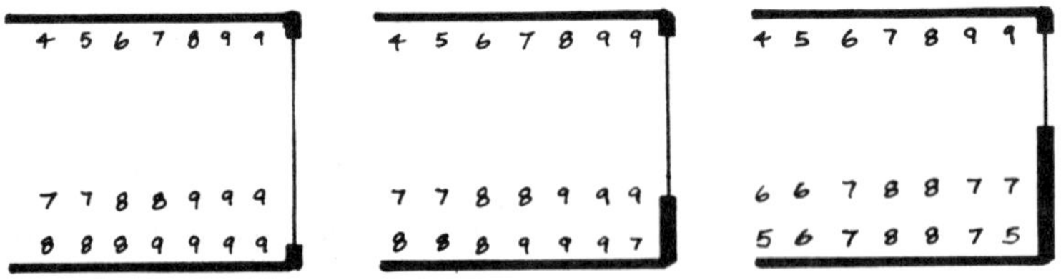

Skylights

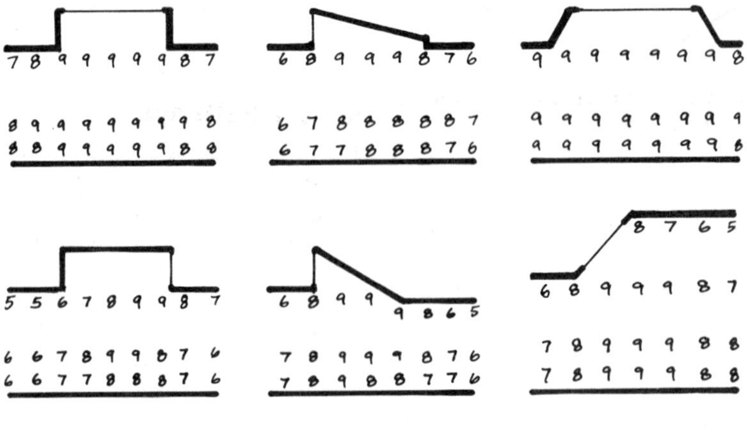

Combinations

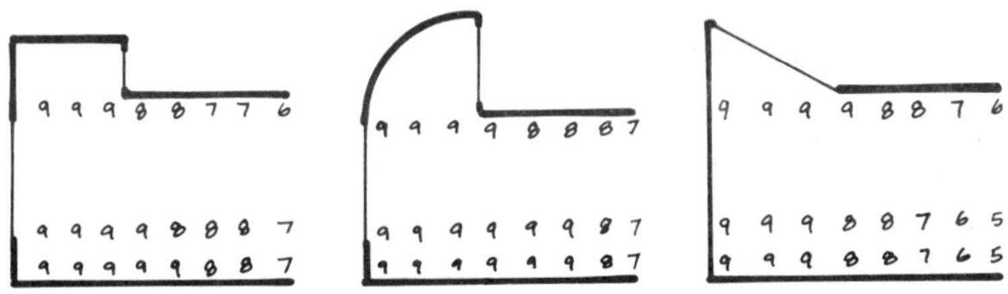

Chapter 9. Solar Access: Planning For Passive Solar Design

1. Introduction

Interest and investment in solar systems for residential uses has sky rocketed in the last two years. What was once the realm of the lone wolf and solar pioneer has now become part of the mass market. The various Federal and State solar tax credits have added further impetus to the rapid transition to solar energy.

Yet one issue still remains unresolved - protection of access to sunshine to enable these solar systems to function in the years ahead. This rather complicated issue will not always be resolved easily or rapidly, particularly in retrofit situations. The increased use of solar energy has many ramifications for the law, planning, and building. At this time, "solar rights" do not exist in much of the U.S., although several bills now under consideration or recently passed will help establish this right. Fortunately however, fairly good protection of solar access can be accomplished using existing land use controls and design practices. Most of these involve primarily the developer and local government staff and their land use planning and building practices and requirements.

This chapter is an introduction to "Solar Access" for developers and local government officials who now bear the responsibility for promoting solar access and includes: An introduction to the basic types of solar systems followed by their requirements for sunshine; The principles of land use planning for solar access; A discussion of the current status of "solar rights" and the possibility of changes in the near future; A solar subdivision in Davis, California, which has applied many of the principles of solar access in its design; and A summary of recent legislation involving solar access.

2. An Introduction To Solar Systems

The energy from the sun and climate resources can and has been used to heat and cool buildings, heat and cool water, generate electricity, cook food, and drive pumps and other mechanical devices. In residential areas, however, the primary near-term uses will be for space heating and cooling and water heating.

Two basic types of solar systems are used for these purposes. Passive solar systems utilize energy from the sun and climate resources to heat and cool without the need for auxiliary energy for pumps or controls. In these space conditioning systems, the windows are the collectors and the elements of the building itself are used for storing the sun's energy for heating or to use the cool night breezes for cooling.

These passive systems are presently the most cost-effective and will soon be the most commonly used. As we have shown even a standard house can be a good passive solar house if it is oriented and designed properly. A recent study demonstrated that, with only minor and inexpensive changes in design a standard house could provide the following performance in the major growth areas of California. Table 1.

The secondmaintype of solar system is the active solar system. These systems collect the sun's energy in one place and then transfer it with pumps, fans, or convection to a separate storage area. A common example of an active system is the flat-plate solar water heating system. Active systems are more expensive than passive solar systems in many cases but are still very competitive with electricity and, in many areas of the U.S., even with gas. They are particularly well suited for heating water in colder areas. A typical system is shown in figure 1.

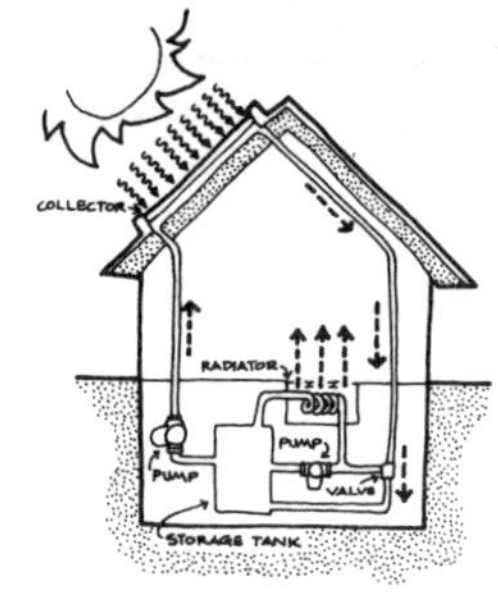

FIGURE 1: An Active Solar System

3. Solar Access Requirements

Both passive and active solar systems require exposure to the sun for heating. In addition, the passive system may need fairly good exposure to the night sky for cooling.

The **Solar Access Requirements** of solar systems may be defined as the necessary field of view required to provide useful heating or cooling. This field of view or skyview will vary depending on four variables:

A. Placement, type, and function of the collector (roof, southwall, space conditioning, water heating, etc.).

B. The yearly use pattern.

C. The daily use pattern.

D. Site and environmental factors.

Although each variable is discussed seperately, they are, of course, interrelated.

A. Collector Position

The sketch below illustrates the primary winter heating skyviews for the four basic collector positions. The ground-mounted collector will be the most difficult to protect as it is the lowest and most susceptible to shading from neighbors to the south. Detached or roof-mounted collectors will usually pose the fewest problems as they are the highest. Generally, the winter sun exposure is the most critical and the most difficult to protect because the sun is very low in the sky. Summer heating skyviews will be easier to protect because the sun is higher in the sky. The summer night sky for radiant cooling is even easier to protect.

B. The Yearly Use Pattern

The yearly use pattern is also an important factor in determining solar access requirements. The three basic uses are:

a) Winter heating only (primarily space heating).

b) All year heating (primarily domestic hot water).

c) Cooling.

The skyview requirements in each of these categories will also depend on the length and intensity of the cold period, the hot period, and the desired percentage of service for the collector system. For example, a solar system which is designed for 100% heating may have much more stringent access requirements than a similar collector which is only expected to meet 80% of the heating demand. The following sections describe some of the solar access considerations for the uses listed above:

a) Winter heating only.

Winter solar space heating is one of the most common and cost-effective uses of solar energy. The solar access requirements for space heating depend on the length and intensity of the cold period, the comfort requirements of the occupants, and the solar design. The length of the heating season, latitude, and microclimate will determine the best type of solar system and the best heating season skyview for a particular area. The nature of the cooling season might also determine the desired type of heating system as an integrated system might be more cost-effective than a "heating only" solar system.

The skyview requirement for space heating will depend on collector position, orientation, use pattern, site, and other factors. The heating season skyview requirements are shown in figure 3 for passive and active solar systems in a relatively clear area at 40°N with a heating season from August 21 to April 21. The passive system skyview is shown with a solid line and the active system skyview is shown with a dotted line. The less costly passive system has a more restrictive skyview than the more expensive active system.

b) Year-Round Heating

Year-round heating for domestic hot water usually requires exposure to the direct sun throughout most of each day all year. The site and type of collector systems can vary widely but the skyview requirement remains virtually the same.

c) Space Cooling

The requirement for natural cooling depends on the length and intensity of the cooling load and the comfort requirements of the occupants. The type of building, the diurnal temperature variation, humidity, and many

	Standard House Facing South with Shading		With Thermal Mass	
San Jose	29	40	47	100
Contra Costa	26	28	68	100
San Diego	47	27	100	86
Orange County	33	27	96	86
Riverside	35	26	96	94
Sacramento	25	26	65	99
Los Angeles	42	25	86	100
	%Heating	**% Cooling**	**%Heating**	**% Cooling**

Table 1. A simple solar house for California.

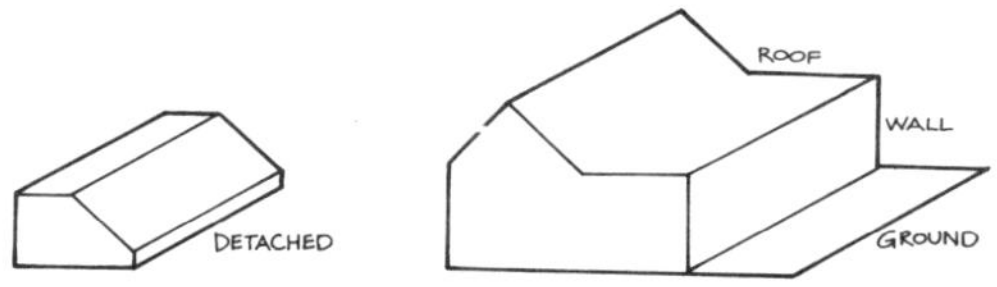

FIGURE 2: Collector Positions

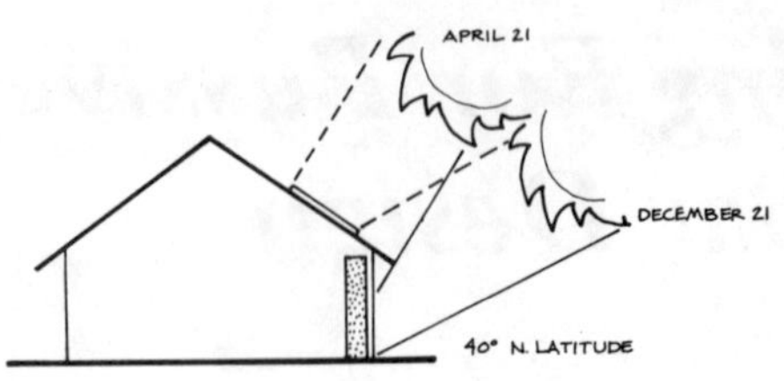

FIGURE 3: Winter Space Heating Skyviews

other factors will determine the type of space cooling that will be used. The most common passive solar systems use cool night breezes for convective cooling in the summer. They may only require solar access in the winter. More sophisticated passive solar systems designed for cooling may use the cold night sky, the cool north sky, evaporation, and convection. The maximum cooling occurs directly overhead at night. The skyview for radiant night sky cooling is shown in figure 4.

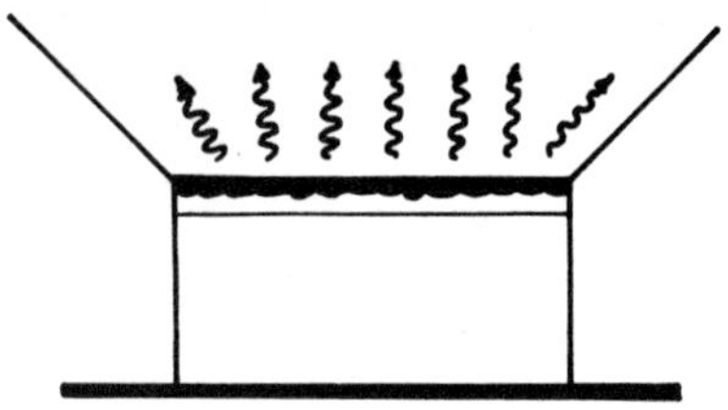

FIGURE 4: Radiant Cooling Skyview

In many areas, both passive space heating and cooling will be desirable. In these areas the skyviews for both heating and cooling must be known and incorporated in planning and design. Protection of exposure to cooling breezes might also be worth considering.

C. Daily Use Pattern

The daily use pattern may also influence solar access requirements. The balance of night vs. daytime temperatures, pattern of available energy, and pattern of needed energy are all important. The collector can be oriented to provide either extra morning heating, balanced heating, or extra afternoon heating. Early morning heating is often desirable in winter in a simple passive house as heat storage from the previous day may be limited and a prompt warming in the morning can be very welcome.

The daily use pattern of a collector can help determine skyview requirements. For example, if large quantities of hot water are desired for clothes washing or showers early in the day, the solar collector for the domestic hot water should be oriented somewhat to the east of south so that the peak collection period comes before noon.

D. Site and Environmental Factors

Many environmental factors may significantly affect solar system performance and, therefore, planning for solar access. These impacts may be either beneficial or adverse. This section details some of the site and environmental factors that may affect solar access.

The basic environmental and site factors which affect solar access include: (a) climate and microclimate, (b) latitude, (c) topography, (d) vegetation, (e) surface type, and (f) man's activities and structures.

a) Climate and Microclimate

The design and construction of a development will create a new microclimate, which can be better or worse than the existing climate depending on the skill and understanding of the designer and his analysis of local microclimatic conditions. A "warm pocket" might require very limited solar access to maintain comfort while an area in a cold drainage channel might require full solar exposure. The Passive Solar Institute will be developing additional materials to aid developers and planners in identifying the favorable and adverse factors for specific local climates.

b) Latitude

The change in the sun's path at different latitudes will, in part, determine solar access requirements. Figure 5 illustrates the difference in altitude and azimuth from the southern to the northern latitudes of the U.S. The northern states have both a greater need for heating and also a more difficult sun to protect - as it is lower in the sky.

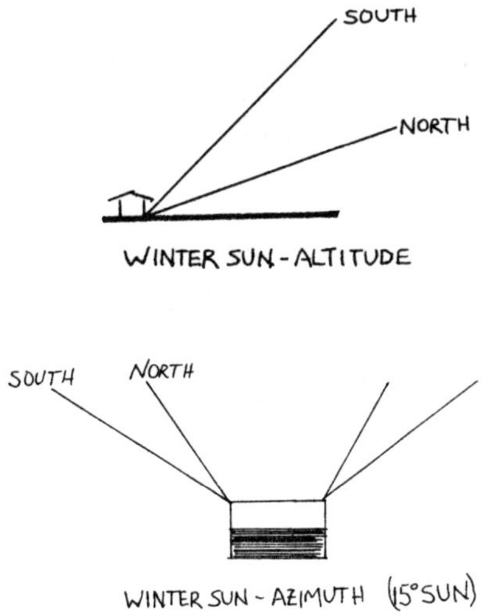

FIGURE 5: Sun Angles and Latitude

c) Topography

The topography of a site exerts a considerable influence on the daily and seasonal patterns of solar radiation. On a steep north facing slope little direct radiation is received at any time during the year, while a south facing slope will consistently receive direct sun. Consequently, cities and developers may wish to put dense housing developments on the south slopes and leave the north slopes for open space or low density development. In steep canyons even the south facing slope may be shaded unless development is carried up the slope.

d) Landscaping

In most areas, vegetation is the most critical factor affecting solar access. Typical residential planning practice in most areas will ensure good solar exposure for roof collectors with any street orientation, except from shading caused by vegetation. Solar exposures for south wall and ground collector systems on E-W streets are usually also protected, except for shading from vegetation. Shading from the bare limbs on deciduous trees can reduce solar radiation by 25 percent or more, and placement of even deciduous trees is very important. Shading from evergreen trees all year and from deciduous trees in the summer can be almost complete and render a solar system inoperable.

Even when the structure is not fully shaded by vegetation, its skyview may be limited. A site with a tree or trees to the south may still have sufficient solar exposure for operation if all of the remaining skyview is used. For example, if a tree to the southwest blocks the afternoon sun, the collector can be oriented to the east. This will shift the skyview to the east as well.

Properly placed trees can provide great savings in both the summer by shading and cooling and in the winter by providing good solar access and reducing infiltration. The proper choices of tree type is also important. Deciduous shade trees should provide full shade in the summer yet have very open branching patterns in the winter. For wind control on the north, evergreen trees may be best. These are shown in figure 6.

FIGURE 6: Planting For Solar Access Can Provide Needed Sun and Shade

f) Man's Impact

Man's planning and construction activities may also have a significant impact on solar exposure and local microclimate. Buildings, for example, affect both solar exposure and wind flow. Well designed streets and buildings can reduce energy use for heating and cooling and increase human comfort. If solar access is ignored in design and planning, structures may be fully shaded and the possibility of using solar systems may be foreclosed, figure 7.

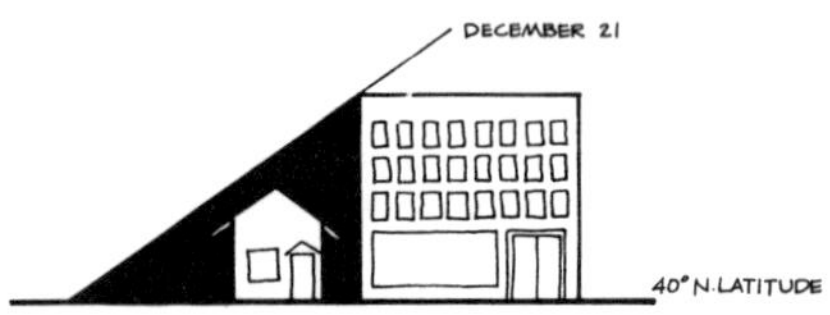

FIGURE 7: Man's Impact On Skyview

4. Planning For Solar Access

The preceding description of solar access requirements should provide a useful introduction to this section on planning for solar access. This section includes a brief description of the critical issues in solar access planning and how they effect development design.

A. Trees

Trees are probably the single most important factor in solar access planning. They are also one of the most difficult factors to control because trees, unlike houses and streets, change dramatically over time and can be planted and grown without review or permit. Yet, trees need not have an adverse impact. They can provide much needed summer shading without blocking collectors if they are planted wisely. Summer shading for houses is needed most on the East and West exposures not the south as might be expected. If trees are planted in the right location, they can provide good shade and full solar exposure in the winter.

Street trees are of particular importance, because they are chosen and maintained by the city. Full, rounded street trees can provide much cooler neighborhoods than small trees, figure 8.

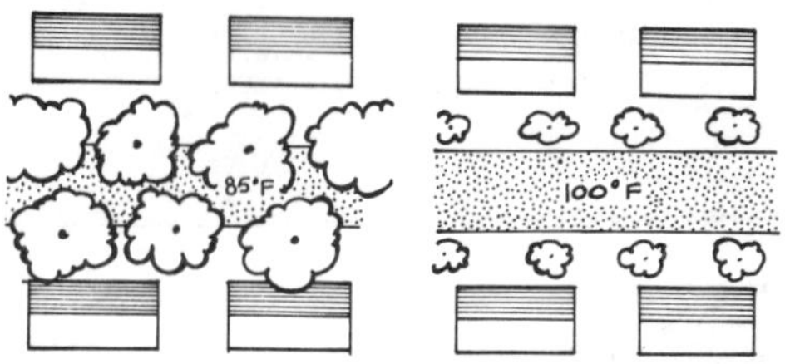

FIGURE 8: Street Trees That Provide Good Shading Can Reduce Neighborhood Temperatures 10-15°F.

The type of tree is important and should receive careful study and consideration. The current policy in most cities has been to select trees which are short, thin, and never drop their leaves. This policy may, in fact, provide the lowest maintenance cost - but it also provides few, if any, other benefits. The savings from full summer shading and solar exposure in the winter with deciduous trees will more than offset the cost of increased maintenance. In some areas, a difference in tree height from the north to south side of the street may be desirable.

Trees and shrubs can also improve ventilation if properly placed. This can be very helpful for cooling houses in areas with cool night breezes. The trees and shrubs act almost like a dam in a river and direct the cooling breeze through the windows for summer cooling, as shown in figure 9.

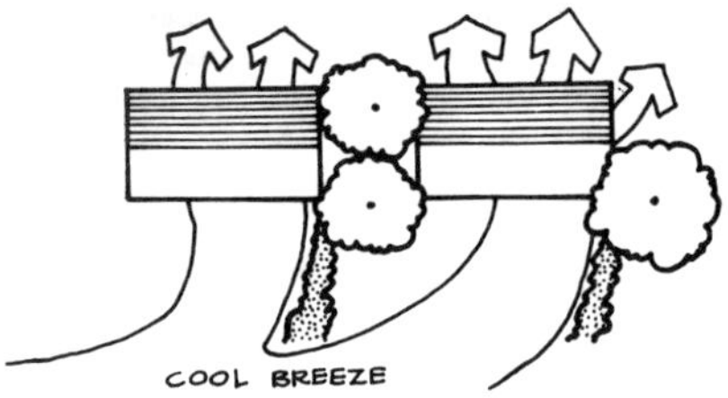

FIGURE 9: Trees and Shrubs Can Help Natural Cooling

B. Street Orientation

In most areas, street orientation is as important as tree placement. Street orientation usually determines house orientation with the houses facing the street and major yards to the front and back. In most areas orienting the street predominantly east-west, with minor bends, is desirable. This orientation ensures that most houses will be simple passive solar houses with lower heating and cooling bills, figure 10. If streets run north-south active solar systems using roof collectors will be easier to use than the less expensive and more reliable passive solar systems, figure 11.

FIGURE 10: Solar Access With East-West Streets and Typical Setback

FIGURE 11: Solar Access on North-South Streets and Typical Setback

In most flat areas, predominantly east-west street direction can be set during development design at no increase in cost. The following plans show how streets were reoriented in a Davis development to meet this goal, figure 12. The number of housing units remained unchanged. In hilly areas this can be more difficult.

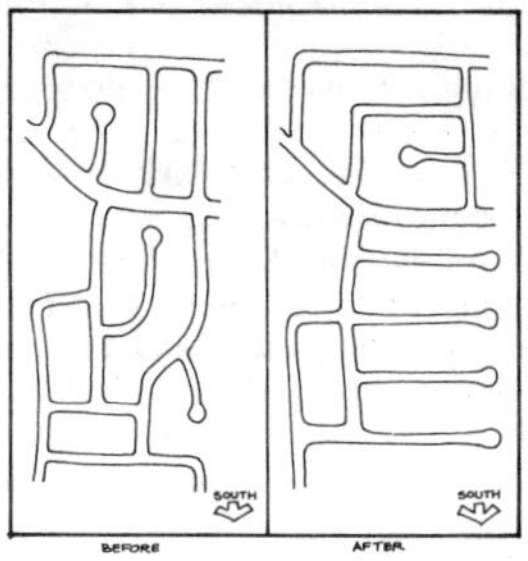

FIGURE 12: A Subdivision Reoriented For Solar Access In Davis, California

C. Setbacks

More flexible setbacks can provide good solar access even where street orientation or vegetation are not favorable for solar utilization. This is shown below for three cases, which are probably the most common cases builders must face, figure 13. Some cities already have "zero" setbacks in some zones and haven't had any problems with them.

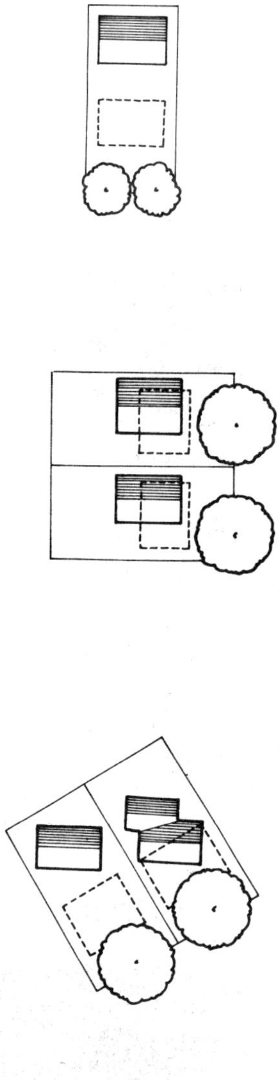

FIGURE 13: Solar Access - Typical Setback Changes Required

D. Fences

In many areas, existing fence regulations may also discourage or prevent homeowners from using passive solar heating and natural cooling. The most common problem is the distance fences must be from the street. For a house with a south facing front yard this fence may block most of the winter sun from the windows. A reasonable reduction in fence setback can provide needed solar access for effective winter heating, figure 14.

FIGURE 14: Fence Setback For Solar Access

E. Height Restrictions

The height restrictions in single family developments usually provide reasonably good solar access; however, when multi-family housing and single family housing are mixed, current restrictions may not protect solar access. Relatively simple changes in design may provide this access at no additional cost. An example is shown in figure 15 for an apartment added to a residential area. The shaded area represents the shadow pattern on December 21 from the apartment building. A berm between the parking lot and the houses might be desirable for noise control.

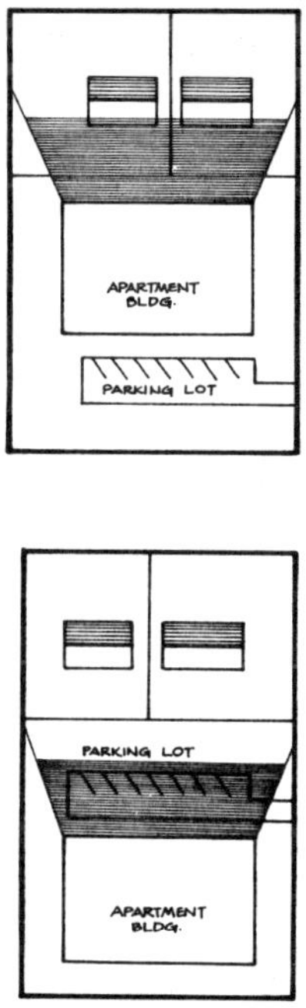

FIGURE 15: Design For Solar Access

Multifamily dwellings offer particularly good solar access for active solar systems because they are tall and well spaced. However, good passive solar access can also be provided with skillful design. Height restrictions may be used to protect good exposure. The following diagrams show how even fairly dense apartments can achieve solar access. The first solar heated condominium is also shown, it was built in 1200 AD.

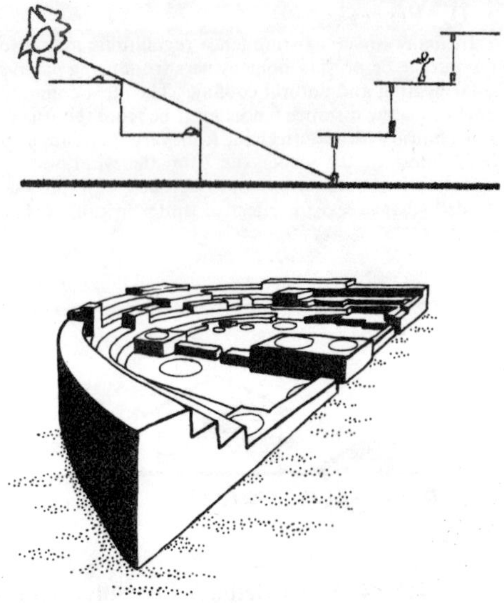

FIGURE 16: Solar Access

F. Street Width

Street width can also be used to improve solar access. Narrower streets allow street trees to be placed further from houses and thus reduce unwanted shading. Narrower streets also provide direct cost savings for the developers and/or local governments for construction and maintenance. In addition, the narrower streets are easier to shade during the summer, figure 17. They are also less costly to build and maintain and safer.

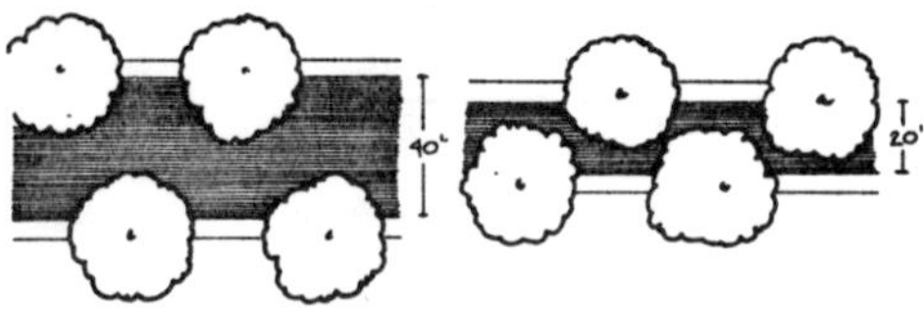

FIGURE 17: Narrow Streets For Solar Access and Natural Cooling

5. Solar Rights

It is important to orient buildings for good solar access for space heating, water heating, and natural cooling even if there is currently no way to ensure that the solar system will not be shaded by neighboring landscaping in future years. Large investments are required for some types of solar heating systems, and a guarantee of "solar rights" in new residential developments should be developed to provide protection for solar home owners.

In California, like most states, the concept of "lights" was invalidated in a California appellate case in 1911. The language of the decision is instructive:

> *"Among the reasons assigned for this view are that this rule (right to light) is not considered to be adapted to existing condition of things in the United States and could not be applied to rapidly growing communities without working mischievious consequences to property owners, and also that in the nature of things there can be no adverse user of light or air . . ."*
>
> —Yuba V. Hinton

Rapid growth is no longer prevalent and as we have already demonstrated with solar heating, the actions of one property owner may indeed, be adverse to another. "When the reason for a law ceases, so too should the law."

In new developments the legal protection of solar access can be established with either easements or covenants. This will be particularly easy in areas where most development occurs in planned unit development. In existing developments, the problem is more difficult because buildings and vegetation may already block the sun. The most reasonable approach here may be simply to carefully review future changes in landscaping and structures in order to determine the best ways to improve solar exposure. Several cities and counties already include solar exposure as an element in their development review and other areas will follow suit. Recently passed legislation concerning solar access is listed in Section 7 of this Chapter.

Another concept which may be worth exploring is the concept of solar zoning. This would include solar exposure in setting the allowable width, height, and depth of development. It would describe a volume of space that would protect solar exposure rather than simply controlling height. One method of doing this is with envelope zoning as shown in figure 18. The American Society of Planning Officials study will provide more explicit examples and specific legal wording in their study of Solar Access.

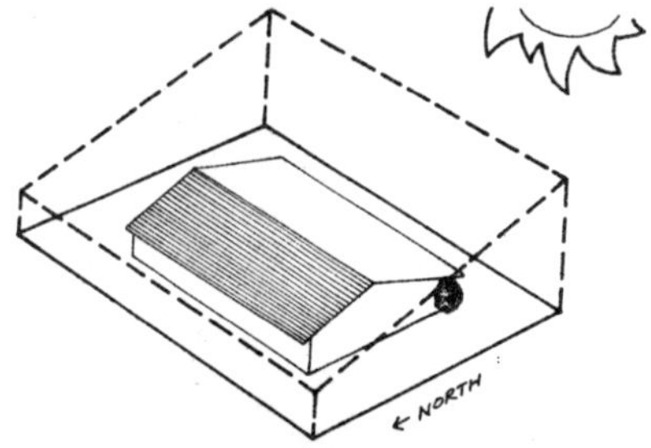

FIGURE 18: Envelope Zoning

The use of the sun is widespread and increasingly valuable, not only for people with solar houses but also for existing buildings with good orientation. As energy prices continue to rise, "solar rights" will be increasingly important and considerable litigation may ensue unless solar access is protected now.

6. A Solar Subdivision

Most of the solar access provisions proposed here are already common practice in several areas of the U.S. (quite often for reasons other than solar access). The best example of a recent solar development in the United States is in California and incorporates most of these considerations. The development is called Village Homes and is located in Davis. The following photo and diagram show both how this solar development is designed and how it looks. All of the houses face south and incorporate basic passive design features. Many include more complete solar systems and achieve 80-90 percent heating and full cooling. George Keck developed passive solar subdivisions near Chicago in the 1940's and the residents there are still reaping the benefits of his vision.

photo by Chad Ankele

FIGURE 19: Village Homes

7. Legislation Involving Solar Access and Solar Rights

Probably no other field of law is changing as rapidly as Solar Access and Solar rights. Many bills have recently passed and many others are still being considered. The following bills are particularly interesting:

Calif: AB 3250: Levine Solar Access - Mandate Broad Local Review, passed 1978, prohibits and cancels legislation preventing use of solar systems.

Calif: AB 2321: Imbrecht, Solar Shade Control Act, prevents tree shading, passed 1978.

New Mexico: HB 294. Makes solar energy a property right, passed 1977.

City of Los Alamos: Zoning Ordinance 173, a clever solar rights mechanism, passed 1977.

In addition, the following list includes many of the jurisdictions involved in Solar Access planning. Contact them for further information.

U.S.: SB 985 (1977) Moakley, did not pass.

States: Arizona - S.B. 1070 (1976), S.B. 1087, (1977); Colorado - S 95 (1975) passed HB 1166 (1976); Florida - HB 3087 (1976); Illinois - SB 693 (1976) HB 1512 (1977) SB 944 (1977) passed; Kansas - HB 2096 (1976) passed; Maryland - HB 360 (1976) passed HB 1024 (1976); Massachusetts - HB 2533 (1977); Minnesota - HB 2064 (1976); North Dakota - HB 1096 (1977); Oregon - HB 2036 (1975) passed; Pennsylvania - HB 728 (1977); Utah - HB 349 (1977); Virginia - HB 2014 (1977).

Local Jurisdictions:

Colorado Springs, Colo., Davis, CA., Indio, CA., Pitkin County, Colo.

In addition to specific laws and regulations many local groups are also setting up policies that encourage solar access and passive solar development. This approach is probably preferable for the near future. Sacramento and San Luis Obispo County planning policies both try to maximize solar development. The Cities of Davis, Indio, and Martinez also encourage developers to maximize southern exposure - in addition to their specific requirements.

The American Society of Planning Officials, The Environmental Law Institute, and Living Systems are preparing materials on Solar Rights and Planning for Solar Access for national distribution. They can provide further information on this topic. Living Systems is preparing a site development manual for Solar Access.

They can be reached at:

American Society of Planning Officials
1313 East 60th
Chicago, IL 60637

The Environmental Law Institute
1346 Connecticut Avenue, NW, Suite 620
Washington, D.C. 20036

Living Systems
Route 1, Box 170
Winters, CA 95694

Further Readings

PLANNING

Amendment to the Uniform Building Code, 1973 Edition, Chapter 53, Energy Conservation and Thermal Insulation. Aspen/Pitkin Planning Department, 130 S. Galena St., Aspen, CO 81611. 1975. 14 pp.

A.S.P.O. (1976) **Energy Efficient Planning, An Annotated Bibliography**, Chicago.

Marty Jaffee *et al* (1978) "Land Use Controls to Protect Solar Access." draft, A.S.P.O.

Bainbridge, D.A. (1975) **Towards an Environmental New Town: A Bibliography**, Council of Planning Librarians, Monticello, Illinois.

Bainbridge, D. and Hammond, J. (1976) **Planning for Energy Conservation**, Living Systems

Bainbridge, D. and Hammond, J. (1976) **Community Design for Energy Conservation**, Living Systems.

Bainbridge, D., Hammond, J. and Kopper, B. (1976) **The Davis Energy Conservation Report**, Living Systems.

Bainbridge, D. and Hunt, M. (1975) "The Davis Experience," **Solar Age**, June

Bainbridge, D. and Kopper, B. (1976) "Energy Conservation for Small Towns," **Small Towns**, V. 7, N. 5, November.

Bainbridge, D. (1977) "Planning for Solar Access: A Glossary," Living Systems for A.S.P.O.

Cramer, R.D. and L.W. Neubauer, (1959) "Solar Radiant Gains Through Directional Glass Exposure," American Society of Heating, Refrigeration and Air Conditioning Engineers, 1958; presented at Lake Placid, New York (June 22-29, 1959) **ASHRAE Transactions**, Vol. 65, p. 409.

Curtis, G.C.S. (1975) **A Design Guide for Residential Areas**, Essex County Council, Tiptree, England.

Eugene, Oregon. **Eugene Code, Section 9.550—Cluster Subdivision in RA and R-1 Districts.** (Ordinance No. 17145, enacted August 12, 1975). Eugene Planning Department, City Hall, Eugene, OR 97401.

Hammond, J., Hunt., M., Neubauer, L. and Cramer, R. (1974) **A Strategy for Energy Conservation**, Living Systems.

Hammond, J., Hunt, M., Neubauer, L. and Cramer, R. (1975) **Davis Energy Conservation Building Code**, Living Systems, Winters.

Hammond, J., Bainbridge, D., Anson, D. (1977) "Skyview Requirements: A Definitional Framework," Living Systems for A.S.P.O.

Hammond, J., Zanetto, J. *et al* (1978) **Planning Solar Neighborhoods**, Living Systems

Harper & Boyle (1976) **Radical Technology**, Random House.

Harwood, C.C. (1978) **Using Land to Save Energy**, Ballinger.

Hunt, M., Bainbridge, D. and Hammond, J. (1977) **Indio Energy Conservation Project**, City of Indio, Living Systems

Hunt., M., Maeda, B., Kopper, B. and Bainbridge, D. (1976) **Davis Energy Conservation Building Code Workbook**, City of Davis

Kopper, W.E. (1974) **Energy Conservation in Davis Households**, UCD, Masters Thesis

McHarg, I. (1969) **Design with Nature**, Doubleday/Natural History.

Meir, R.L. (1974) **Planning for an Urban World**

Miller, A., Hayes, G. and Thompson, G. (1977) **Solar Access and Land Use**, NSHCIC, Rockville, Md.

Myhra, David (1974) "Energy Husbandry in the Home Building Industry—A Look at a Planned Unit Development." **Public Utilities Fortnightly**. Public Utilities Reports, Inc., 1828 L St., N.W. Washington, D.C. 20036. March.

Neubauer, L.W. (1972) "Optimum Alleviation of Solar Stress on Model Buildings," **Transactions of the American Society of Agricultural Engineers**, Vol. 15, No. 1, pp. 129-132.

New York Legislative Commission on Energy Systems (1977) **A Forum on Solar Access**, NSHCIC, Rockville, Md.

Olgyay, V. (1963) **Design with Climate**, Princeton Press

Pleijel, G. (1954) **The Computation of Natural Radiation in Architecture and Town Planning**, Statens Namnd for Byggerandstorsking, Stockholm.

Real Estate Research Corp. (1974) **The Cost of Sprawl**, USGPO, Washington, D.C.

Ridgeway, J. (1977) **The Davis Experiment**, The Elements, 1747 Connecticut Avenue, Washington, D.C. 20009. ($2.00)

Safdie, M. (1975) **For Everyone A Garden**, MIT Press.

Stein, J. (1976) **Energy and Architecture: Implications for Human Settlements**, I.I.E.D.

Washington, D.C., Area. (1975) **Energy, Land Use, and Growth Policy: Implications for Metropolitan Washington.** Second ed. Metropolitan Washington Council of Governments, 1225 Connecticut Ave., N.W., Washington, D.C. 20036.

Witherspoon, R., and Jon A. Abbett, and Robert Gladstone. (1976) **Mixed Use Developments: New Ways of Land Use**, Urban Land Institute, Washington, D.C.

Jan Hamrin (1978) **Solar Energy Conserving Communities: Implications for Public Policy**, Doctoral Thesis, UC Davis

LANDSCAPE

Bainbridge, D.A. (1976) **Planning for Energy Conservation**, Living Systems, Winters, Calif.

Bernatsky, A. (1966) "Climatic Influences of Greens and City Planning," **Anthos** v. 1, n. 1

Deering, R.B. (1953) "The Importance of Microclimatic Problems in Garden Design," **The National Horticultural Magazine** (October) pp. 226-230.

Deering, R.B. and Brooks, F.A. (1954) "The Effect of Plant Material Upon the Microclimate of House and Garden," **National Horticultural Magazine**, July.

Deering, R.B. (1955) "Effective Use of Living Shade," **California Agriculture** (September), pp. 10, 11 & 15.

Everson, G.J., Neubauer, L.W., and Deering, R.N. (1956) "Environmental Influence on Orientation and House Design to Improve Living Comfort," **Journal of Home Economics**, v. 48, n. 3, March.

Ferber, A.E. (1969) **Windbreaks for Conservation**, USDA, S.C.S.

Geiger, R. (1950) **The Climate Near the Ground**, Cambridge University.

Hammond, Hunt, Neubauer, Cramer (1974) **A Strategy for Energy Conservation**, Living Systems.

Landsberg, H. (1947) "Microclimatology for Planners," **Architectural Forum**, March.

Myrup, L.O. and Morgan, D.L. (1972) **Numerical Model of the Urban Atmosphere: The City-Surface Interface**, V. I. Contributions in Atmospheric Science, Number 4, University of California, Davis.

Neubauer, L.N. (1970) "Optimum Alleviation of Solar Heat Stress on Model Buildings," ASAE paper 70-401.

Olgyay, V. (1963) **Design with Climate**, Princeton University.

Robinette, G.O. (1972) **Plants, People and Environmental Quality**, U.S.G.P.O.

White, R.F. (945) **Effects of Landscape Development on Natural Ventilation**, RR 45, Texas Engineering Experiment Station, Austin.

Robinette, G.O. (1978) **Landscape Design for Energy Conservation**, E.D.I., Reston, Va.

SOLAR RIGHTS

Bainbridge, D. (1975) "Solar Rights," Testimony before Assembly Judiciary Committee, Living Systems, Winters.

Bainbridge, D. (1979) "Solar Rights: The New Legal Frontier," **New Energy**, in press.

Bules, D. (1978) "Solar Rights," Solar Office, Energy Commission.

"Clawson v. Primrose," 4 Del. Ch. 643.

Cox, H. (1871) **The Law and Science of Ancient Lights.**

Environmental Law Institute - major study underway.

Figueroa, D. (1976) "EIR Energy Provisions," Davis Planning Department.

Harris, W.R. (1975) "Is the Right to Light a California Necessity," RAND.

Hendo, H. (1971) "Right to Sunshine and Building Standards Law," **Juristo**, 481, 55.

Jsemon V. Bradford Corporation (1922) **Law Journal**, 91 Chancery Division.

Swarbricks, J. (1930) **Easements of Light**, Batsford.

Thomas, W.A., Miller, A.S, and Robbins, R.L. (1978) **Overcoming Legal Uncertainties About the Use of Solar Systems**, American Bar Foundation.

Wilson, Jones, Morton, and Lynch (1976) "A Model Solar Utilization Ordinance," Santa Clara/HUD.

The Solar Law Reporter, Solar Energy Research Institute, 1536 Cole Blvd., Golden, Colo. 80401.

TRANSPORTATION

Bainbridge, D.A. (1974) **Bikeway, Planning and Design**, DABEP, Davis.

Curtis, G.C.S. (1975) **A Design Guide for Residential Areas**, Essex County Council, Tiptree, London.

Hirst, Eric. (1973) "Transportation Energy Use and Conservation Potential," **Bulletin of the Atomic Scientists**, Vol. 29, No. 9 (November). pp. 36-42.

Public Works Dept. (1975) "Development of Bicycle and Street Facilities in Davis," Davis, CA.

Chapter 10: Build It Yourself

One of the best features of passive solar systems is their simplicity. Almost anyone can build or remodel a house for passive solar heating and cooling, daylighting, and hot water heating. The tools and materials needed are relatively inexpensive and only basic construction knowledge is needed for most projects.

The easiest way to begin may be with simple projects like insulated drapes or shutters and exterior shades and arbors. With success here you may wish to try a more complicated project - like adding a south window or thermal mass storage in front of your south windows. It's simple!

Check these things carefully - structural integrity, weatherizing, water-proofing (flash everything carefully), wind load, fire safety (exposed plastic foam is a poor idea), and seismic safety. Build things carefully and they will last a long time. As my friend and building advisor Denny says, "Build for elephants, 100 mile per hour winds, torrential rains, and a hundred years."

A passive solar water heater is also a good project for the beginner. It requires only basic carpentry and plumbing skills when built on the ground or on a flat roof. Building one into a roof is a little bit trickier.

After you've undertaken a series of projects like these you'll probably want to design and build your own passive solar house. There is no reason you shouldn't. Become familiar with the skills needed by working with a small builder, taking an owner-builder class, or hire a contractor to oversee you.

You can either pay for the house yourself or get a loan from the bank. You should at least consider paying for it yourself rather than financing it. You can build a small core and then add on as your finances and time allow. This may seem difficult at times - but you may find there are many rewards in saving yourself from house payments.

If you decide to finance your house, here are the steps you'll usually take: 1) find a lot with good solar exposure, 2) buy or option it, 3) design your house, 4) calculate performance, 5)optional: have an established firm check your plans and calculations, 6) get building department approvals, 7) go to the bank (banks) and obtain a construction loan using the lot as collateral. If your house is unusual (underground or trombe wall or . . .) then make sure you've got everything calculated, described, and approved. An engineer's signoff will often appease even the most uncertain bank officer, 8) build the house (or have it built), 9) finish the house and get building department approval, 10) refinance to a regular loan, and 11) enjoy your comfortable and economical solar home.

This is it in a nutshell, for further information I recommend the following books:

Les Scher (1974) **Finding and Buying your Place in the Country**, Collier.

Ken Kern (1975) **The Owner Built Home**, Chas. Scribners. Excellent overview and coverage of a range of materials and technology.

Barbara and Ken Kern (1977) **The Owner Built Homestead**, Chas. Scribner. More broad than the Home book with shop and farm operations as well. Delightful.

Ken Kern, Ted Kogan, Rob Thallon (1976) **The Owner Builder and the Code**, Owner Builder Publications. How to build a good, simple house despite the codes.

Navpers (1972) **Basic Construction Techniques for Houses and Small Buildings**, Dover. This navy training manual is a good book on basic construction. It also has an excellent chapter on logging operations and heavy construction.

L.O. Anderson (1971) **Woodframe House Construction**, Craftsmen Books, Solana Beach. More information on simple woodframe construction than Navpers. Very clearly described and well illustrated.

H.P. Richter (1968) **Wiring Simplified**, Park Publishing Co. Short and to the point. It's the only one I use.

Albert Dietz (1977) **Dwelling House Contruction**, The MIT Press. This book is written almost like a textbook. It covers all phases of house construction except plumbing and wiring.

Robert McHugh (1977) **Working Drawing Handbook**, Van Nostrand. This book will get you through plan drawing. You will probably be surprised how easy it is.

Rex Roberts (1964) **Your Engineered House**, M. Evans. This bo k has one of the best introductions on the Why, What, Where and When of house building. Unfortunately the How is flawed. Buy the book, read chapters 1, 2, 3, and 5 and 6. Rely on the other books for How. Rex almost completely disregards energy.

Jay Baldwin and Stewart Brand (1978) **Soft-Tech**, Penguin. Worth getting for a number of reasons but for house building read Jay's "One Highly Evolved Toolbox." It will save you money, frustration, and quite possibly your life.

Grayham Blackburn (1974) **Illustrated Housebuilding**, Viking Press. I'm including this on Jay's recommendation. He says it is just right for a novice and still useful for an experienced builder.

The Editors (1973) **The Complete Do-It-Yourself Manual**, Readers Digest. This will help you do your plumbing, wiring, etc. Descriptions and illustrations are very well done.

That should do it. For specific information on building details for different passive systems look at the reference lists and materials in the appropriate sections of the catalog. The following groups offer instruction in owner-building and are all recommended:

Shelter Institute
72 Front St.
Bath, ME 04530

Cornerstones
54 Cumberland St.
Brunswick, ME 04011

The Owner Builder Center
269 Cardinal Road
Mill Valley, CA 94941
415 - 383-6924

The Habitat Center
573 Mission St.
San Francisco, CA
95105

You may also find the following books useful:

Inspiration

Stewart Brand and Friends (1972) **The Last Whole Earth Catalog**, Co.Ev., Sausalito

Lloyd Kahn (1973) **Shelter**, Shelter Publications, Bolinas

Art Boericke and Barry Shapiro (1973) **Handmade Houses**, Scrimshaw Press

A. Wade and N. Ewenstein (1977) **30 Energy Efficient Houses You Can Build**, Rodale Press

Dave Bainbridge, John Hofacre and Judy Corbett (1979) **Village Homes: Solar House Designs**, Rodale Press

Norma Skurka and Jon Naar (1976) **Design for a Limited Planet**, Ballantine

Hassan Fathey (1973) **Architecture for the Poor**, Univ. of Chicago Press

Bernard Rudofsky (1964) **Architecture without Architects**

Christopher Williams (1974) **Craftsmen of Necessity**, Vantage

Design

Carol Stoner (1977) **Goodbye to the Flush Toilet**, Rodale Press

Alexander Kira (1976) **The Bathroom**, Viking Press

Murray Milne (1976) **Residential Water Conservation**, Calif. Water Resources Center, Report #35

Witold Rybznyski (1974) **The Ecol Operation**, McGill University, Canada

Anthony Adams (1977) **Your Energy Efficient House**, Garden Way, Charlotte, Vt.

Brenda Vale (1974) **The Autonomous House**, Thames and Hudson

Etienne Grandjeanne (1973) **Ergonomics of the Home**, Halstead Press

Building

Karl and Sue Schwenke (1976) **Build Your Own Stone House**, Garden Way, Charlotte, Vt.

Merilees and Loveday, **Pole Building**

Eugenie Wallace and Stewart Elliott (1977) **The Timber Framing Book**

Charles Neal (1978) **Do-It-Yourself Housebuilding**

Charles Wing and John Cole (1976) **From the Ground Up**, Cornerstone

Time-Life Series, excellent primers on Basic Wiring, Plumbing, Weatherproofing, etc.

Sunset series, excellent primers on Basic Wiring, Basic Plumbing, Remodeling, Kitchens, etc.

Al Burns (1976) **The Skylight Book**, The Running Press

Remodelling

Donald Brann (1974) **How to Rehabilite Abandoned Buildings**, Directions Simplified

Gerald Sherwood (1975) **New Life for Old Dwellings**, U.S. Forest Service, Ag. Handbook 481.

The Habitat Center, photo: Lynn Nelson

Chapter 11: Access to Components and Consultants

1: Components

This chapter includes listings of passive solar components and consultants. The components are listed first and are divided in 17 different categories. The consultants for this first issue are divided into just two categories, passive design and daylighting, more specific listings will be used in the Second Catalog to help you more easily identify a consultant to meet your needs.

1. Passive Solar Components

The following listing of passive solar components has been made as complete as time allowed; however, some products are probably still not listed. Please let us know if we missed you—or if you know of anyone making a product we should list.

A listing here in no way constitutes a recommendation or warranty and you should evaluate product claims and specifications carefully to make sure they will work in your system.

Many of these companies are small and struggling to get established. Therefore, when asking for information send at least a self-addressed, stamped envelope (8½" x 11") and/or a check for $2 to $5. And only ask for information if you are serious about it. The Whole Earth Catalog inadvertently put several companies out of business with requests for information that were never followed with orders. So when you see a product that you can/should use don't just think about it—order it. And be patient if it takes a while to get there.

If you have the means to do so, consider becoming a dealer or distributor for passive solar products. Many of these products need regional representatives—and will pay handsome returns to those who invest now.

The Passive Solar Institute will be issuing a series of market reports concerning passive solar products. These will discuss the principles, design considerations, and market estimates for passive products not currently available—or available in limited supply. These will be published irregularly during the year. Tentatively planned reports include:

TR - 1-79: Water Wall Modules
TR - 2-79: Passive Solar Hot Water Heaters
TR - 3-79: Insulated Shutters
TR - 4-79: Insulated Drapes
TR - 5-79: Indirect Evaporative Cooling
TR - 6-79: Phase Change Materials in Passive Design

Write us concerning availability and cost. We are also interested in helping with more complete product development.

The classification of products roughly follows the order of the book and chapter on solar design. Your comments on the order and classifications would be appreciated.

Companies printed in bold type responded to the letters we sent out. Companies in light type either didn't respond or were suggested by others too late for confirmation.

Tell them you learned about them in the Passive Solar Catalog!

A. Index to Components

1. Glazing
- 1.1 Glazing materials
- 1.2 Windows
- 1.3 Skylights
- 1.4 Greenhouses
- 1.5 Other

Glazing Control
- 2.1 Glazing Control materials
- 2.2 Insulated drapes
- 2.3 Insulated shutters

3. Solar Control
- 3.1 Materials
- 3.2 Movable shades (see 2.3)
- 3.3 Awnings
- 3.4 Other

4. Reflectors

5. Thermal Mass
- 5.1 Water tanks and containers
- 5.2 Masonry
- 5.3 Phase change materials
- 5.4 Other

6. Insulation
- 6.1 Building
- 6.2 Pipes
- 6.3 People

7. Weatherizing
- 7.1 Caulking
- 7.2 Weatherstripping
- 7.3 Other

8. Passive Space Conditioning Systems

9. Passive Hot Water Heaters
- 9.1 Tanks
- 9.2 Valves & etc.
- 9.3 Other

10. Daylighting

11. Solar Electric
- 11.1 Wind
- 11.2 Photo Voltaic
- 11.3 Solar Thermal
- 11.4 Other

12. Backup Systems
- 12.1 Fans and Blowers
- 12.2 Ceiling Fans
- 12.3 Ventilation/Heat Exchangers
- 12.4 Vents
- 12.5 Space Heating
- 12.6 Space Cooling
- 12.7 Water Heating

13. Appliances
- 13.1 Refrigerators, Stoves, etc.
- 13.2 Toilets
- 13.3 Water conservation

14 Instruments

15. Living the Good Life

16. Solar Advocates and Associations
- 16.1 Associations
- 16.2 Lobbies

17. Information
- 17.7 Magazines
- 17.2 Information services and books
- 17.3 Product Information

18. Simulation

photo : Susan Maas Paul ©1978
Alternative Source of Energy

1. Glazing

1.1 Glazing Materials

CY/RO Industries
Wayne, NJ 07470

DuPont Co.
Plastic and Resins Dept.
Wilmington, DE 19898

Kalwall Corp.-Sunwall/Sunlite
P.O. Box 237
Manchester, NH 03103

Lasco Industries
3255 East Miraloma Ave.
Anaheim, CA 92806

Lasco Industries
8015 Dixon Drive
Florence, KY 41042

ASG Industries Inc.
P.O. Box 929
Kingsport, TN 37662

J.W. Carroll & Sons
22600 S. Bonita Street
Wilmington, CA 90744

Chemplast Inc.
150 Dey Road
Wayne, NJ 07470

Lupton Div. of Conalco
700 East Godfrey Avenue
Philadelphia, PA 19124

National Ind. Sales, Inc.
6651 West 99th Street
Chicago Ridge, IL 60415

3M Company
3M Center, Bldg. 223-2
St. Paul, MN 55101

Rohm and Haas
Independence Hall West
Philadelphia, PA 19105

Structured Sheets Inc.
196 E. Camp Avenue
Merrick, NY 11566

Vistron Corp.
Filon Division
2333 South Van Ness Ave.
Hawthorne, CA 90250

Solar Roll
Bio Energy Systems
Mountaindale Rd.
Spring Glen, NY 12483

1.2 Windows

Anderson Corporation
P.O. Box 12
Bayport, MN 55003

Alcoa
600X Alcoa Bldg.
Pittsburgh, PA 15219

Caradco Window
P.O. Box 658
Dubuque, Iowa 52001

Blombert Window Systems
1415 Blair Avenue
Sacramento, CA

Pella Windows and Doors
100 Main Street
Pella, Iowa 50219

PPG Retrofit Window System
P.O. Box 2811
Pittsburgh, PA 15230

Plaskolite
P.O. Box 1497
Columbus, OH 43216

1.3 Skylights

Aluminex Skylights
2408 Forney
Los Angeles, CA 90031

Bristol Fiberlight
3200 Halliday
Santa Ana, CA 92705

Roto-Roof Window
P.O. Box 758
Essex, Conn. 06426

McKure Metal Products
570 Bryant
San Francisco, CA 94107

O'Keefes, Inc.
75 Williams Avenue
San Francisco, CA 94124

Pfeiffer Skylight Corp.
14692 Wicks Blvd.
San Leandro, CA 94577

Rollamatic Roofs
1400 Yosemite
San Francisco, CA 94124

Ventarama Skylight Corp.
40 Haven Avenue
Pt. Washington, N.Y. 11050

Wasco Products
P.O. Box 351
Sanford, Me. 04073

Naturalite Inc.
3233 West Kinsley Rd.
Garland, TX 75228

1.4 Greenhouses

Energy Factory
5622 E. Westover, Suite 105
Fresno, CA 93727

Four Seasons Solar Products
672 Sunrise Highway
West Babylon, NY 11704

Janco Greenhouses
9390 Davis Ave.
Laurel, MD 20810

Lord and Burnham
Main Street
Irvington, NY 10533

McGregor Greenhouse
P.O. Box 36
Santa Cruz, CA 95063

Pete Reimuller - The Greenhouseman
980 17th Avenue
Santa Cruz, CA 95062

Santa Barbara Greenhouses
390 Dawson Drive
Camarillo, CA 93010

Solar Collectors of Santa Cruz
2902 Glen Canyon Road
Santa Cruz, CA 95060

Solar Room
P.O. Box 1377
Taos, NM 87571

Solar Technology Corp.
2160 Clay Street
Denver, Colo. 80211

Sturdi Built Mfg. Co.
11304 S.W. Boones Ferry Road
Portland, OR 97219

Texas Greenhouse Co.
2709 St. Louis Ave.
Fort Worth, TX 76110

Vegetable Factory Dept. S.C.
P.O. Box 2235
Grand Central Station
New York, NY 10017

1.5 Other

Passive Lifters
Dalen Prod. Inc.
201 Sherlake Dr.
Knoxville, Tenn. 37922

Solarama
P.O. Box 210
Colorado City, AZ 86021

Heat Motors Inc.
635 W. Grandview Ave.
Sierra Madre, CA 91024

2. Glazing Control

2.1 Glazing Control Materials

A. Shutter Material

Metro Supply
1420 47th
Sacramento, CA 95822

Celotex Corp.
1500 North Dole Mabuy Ave.
Tampa, FL 33607

B. Drape Material

Dacron Holofil II
E.I. DuPont DeNemours & Co.
Wilmington, DE 19898

Polarguard
Fiber Industries Inc.
Barclay Downs Drive
Charlotte, NC 28209

Space Blanket
Thermos Division
King Seeley
Norwich, CT 06360

Foylon - Reflective Material
Duracote
350 North Diamond
Ravenna, Ohio 44266

Thinsulate - Insulating Mesh
223-6SW
3M Center
Saint Paul, MN 55101

2.2 Insulated Drapes

Appropriate Technology Corp.
P.O. Box 975
Battleboro, VT 05301

Drapery Solar Liner
Sol-Pub
P.O. Box 2351
Gaithersburg, MD 20760

Insulating Shade Company
P.O. Box 406
Guilford, CT 06437

Thermal Technology Corp.
P.O. Box 130
Snowmass, CO 81654

Wind-n-Sun Shield
P.O. Box 1434 MG
Melbourne, Florida 32935

2.3 Insulated Shutters

Shutters, Inc.
110 East 5th Street
Hastings, MN 55033

Amrol Exterior Rolling Shutters
Pease Company
New Castle, IN

Ark-tic-seal Systems, Inc.
P.O. Box 428
Butler, Wi. 53007

The Solar Insulation Panel
Sol Pub
P.O. Box 2351
Gaithersburg, Md. 20760

Pease Co. - Ever Straight
7100 Dixie Highway
Fairfield, Ohio 45023

Roll-A-Way Shutters, Inc.
393105 Terrace N.E.
St. Petersburg, Fla. 33702

Rolenco
Suite 103, I.E. Tower
Cedar Rapids, Iowa 52401

Rolscreen Company
Pella, IA 50219

Shutter Shield
C.C. Davidson & Assoc.
1014 Baldwin Avenue
Pontiac, MI 48057

ThermaRoll Corp.
512 Orchard Street
Golden, Colo. 80401

ThermaRoll, Suite 209
4600 Post Oak Place Drive
Houston, TX 77027

Movable Insulation Systems
Zomeworks
P.O. Box 712
Albuquerque, NM 87013

Serrande of Italy
P.O. Box 1034
West Sacramento, CA 9569

Solar Energy Components Inc
212 Welsh Pool Rd.
Lionville, PA 19353

3. Solar Control

3.1 Solar Control Materials

A. Metal Shadescreen:

Koolshade Corporation
722 Genevieve Street
Solana Beach, CA 92075

Kaiser Shadescreen
ASC Distributing, Inc.
1645 West Buckeye
Phoenix, AZ 85007

B. Fiberglass Shadescreen:

Phifer Wire Products, Inc.
2612 Irving Boulevard
Dallas, Texas 75207
Charles Kimberl, District Manager
214/630-9050

Phifer Wire Products, Inc.
Post Office Box 1700
Tuscaloosa, Alabama 35401
CALL TOLL FREE
1-800/635-5955

Fiber Glass Div. Dept. 92
J.P. Stevens Co.
P.O. Box 1138
Walterboro, South Carolina 29488

Owens Corning Fiberglass
Granville OH 43023

Vimco Solar Shield
P.O. Box 212
Laurel, VA 23060

PPG Industries Inc., Suite 302
300 North Lee Street
Alexandria, VA 22314

Chicopee Manufacturing Company
Cornelia, GA 30531

Phifer Western
14408 East Nelson
City of Industry, CA 91744

3.2 Movable Shades

See 2.3 Shutters

3.3 Awnings

The Solar Protection Co.
Box 98
New Canaan, Conn. 06940

Canvas Products Association
350 Endicott Bldg.
St. Paul, Minn. 55101

Sunkit Fabrics
12937 Venice Blvd.
Los Angeles, CA 90066

Sunkit Fabrics
Crooked Lane, Rt. 1
Columbia City, Ind. 46725

3.4 Films

American Solar Control Co.
5098 Gazania Drive
San Jose, CA 95111

3M Energy Control Products
1121 Grandview Drive
So. San Francisco, CA 94080

General Solar Corp.
451 S. Stonestreet Ave.
Rockville, Md. 20850

4 Reflective Materials

Edmund Scientific
7780 Edscorp Bldg.
Barrington, NJ 08007

Solar Usage Now
Box 306
Bascomb, Ohio 44809

5. Thermal Storage

5.1 Water Tanks and Containers

A. Special Tanks

Kalwall Corporation
Solar Components Division
P.O. Box 237
Manchester, N.H. 03103

One Design, Inc.
Mountain Falls Route
Winchester, VA 22601

Tabline Company (custom only)
P.O. Box 1135
Berkeley, CA 94701

B. Culverts

Armco Steel - Culverts
925 Village Center
Lafayette, CA 94549

C. Drums

Western Drums
128 Harbor Way
South San Francisco, CA

A.W. Sorich Co.
429A South Canal
South San Francisco, CA

D. Cans

Bernarr Jonish-Container Div.
S. Riekes/Son
1818 Leavenworth St.
Omaha, Neb. 68102

American Can
100 Park Ave.
N.Y. N.Y. 10017

E. Plastic Drums

Advanced Chemical Technology
1100 South Azusa
City of Industry, CA
91744

F. Other

Doerr Metal Products
Box 10
Larned, KA 67550

5.2 Masonry

A. Paint

Epoxy Fortified Floor Paint
Fuller-O'Brien
South San Francisco, CA 94080

B. Texture

Indesco
1929 Cross Highway
Fairfield, Conn. 06430

C. Sources

Intl. Masonry Institute
823 15th St. N.W. Suite 1001
Washington, D.C. 20005

National Concrete Masonry Association
P.O. 135
McLean, VA 22101

5.3 Phase Change Materials

Calmac Mfg. Corp.
P.O. Box 710 E
Englewood, NJ 97631

McKim Solar Energy Inc.
2627 E. Admiral Place
Tulsa, OK 74110

Oem Products, Inc.
2701 Adamo Drive
Tampa, FL 33605

Solar Mkt.
P.O. 1776
Omaha, Neb. 68101

Solar Shingles Company
14532 Vanowen Street
Van Nuys, CA 91405

Solar Shelter Engin. Co.
P.O. Box 179
Kutztown, PA 19730

Architectural Research Corporation
Sol-ar-tile
13030 Wayne Road
Livonia, MI 48150

American Solarize, Inc.
P.O. Box 15
Martinsville, NJ 08836

Sun-Heet Inc.
2624 South Zuni
Elglewood, CA 80110

Solar Inc.
P.O. Box 246
Mead, NE 68041

5.4 Other

Burke Rubber Hypalon
2250 South 10th
San Jose, CA

Gates Engineering-Sealants
100 S. West Street
Wilmington, Deleware 19899

6. Insulation

6.1 Building Insulation

Red Cedar Shingle and Handsplit Shake Bureau
Suite 275
515 - 116th Ave. NE
Bellevue, WA 98004

Plasticool (coating-insul.)
Coating Labs
3133 E. Admiral Place
Tulsa, Oklahoma 74110

BASF Wyandotte Corp.
Styropor Division
1609 Biddle Ave.
Wyandotte, Mich 48192

Johns-Mansville
2600 Campus Drive
San Mateo, CA 94403

R Max Inc. - Ext. Sheath Insul.
13524 Welch Rd.
Dallas, Texas 75240

Home Foamers
Sci. Applications Inc.
P.O. 619
Mt. Pleasant, Iowa 52641

6.2 Pipe Insulation

Teledyne Mono-Thane
1460 Industrial Parkway
Akron, Ohio 44310

Urethane Moulding Inc.
RFD #3 Rt. 11
Laconia, NH 03246

Energy House
P.O. 5288
Salem, Ore. 97304

Solar Usage Now
Box 306
Bascom, Ohio 44804

6.3 People Insulation

Damart
1811 Woodbury Ave.
Portsmouth, NH 03805

Thermoply Insulation
P.O. Box 10
3000 W. Beecher Road
Adrian, Mich. 49221

Olam - Pile garments
P.O. 460
Spring Hope, NC 27882

North Face
1234 Fifth Street
Berkeley, CA 94710

Frostline Kits
Frostline Circle
Denver, CO 80241

Early Winters
110 Prefontaine Place South
Seattle, Washington 98104

R.E.I.
P.O. Box 88125
Seattle, WA 98188

Sierra Design
247 Fourth St.
Oakland, CA 94607

L.L. Beans
Freeport, ME 04033

Eastern Mountain Sports
One Vose Farm Road
Peterborough, NH 03458

7. Weatherizing

7.1 Caulking

Insta-Foam Inc.
Joliet, Il. 60435

Silicone Caulk
Silicone Products Dept.
General Electric
Waterford, N.Y. 12188

Energy House
P.O. 5288
Salem, Oregon 97304

Sears
Wards
local hardware store

7.2 Weatherstripping

Energy House
P.O. 5288
Salem Oregon 97304

Sears
Wards
local hardware store

7.3 Other see also windows

8. Passive Space Conditioning Systems

Solar Staircase, PSI
Norman Saunders
15 Ellis
Weston, MA 02193

Passive Solar-Variant Homes
Savell
575 Birch Ct, Suite A
Colton, CA 92324

Green Mountain Homes $2.25
Royalton, Vermont 05068

Skytherm
2424 Wilshire Blvd.
Los Angeles, CA 90057

Advance Cooler Mfg. Co.
Clifton Park, N.Y. 12065

One Design Inc.
Mountain Falls Rt.
Windrester, VA 22601

Kalwall
P.O. 237
Manchester, NH 03103

Pittenger Roof System
Tempe, Arizona

Mother Earth News - #84010 Underground House Plans $135
P.O. Box 70
Hendersonville, NC 28739

9. Passive Water Heaters

9.1 Elec. tank liners:

American Appliance
Santa Monica, CA

Rheem (may make available in midwest)
7600 S. Kedzie
Chicago, Illinois 60652

9.2 Drain Valve - Vacuum breaker air bleed valve

Richdell Inc.
Solar Division
P.O. Drawer A
Carson City, CA 89701

Energy House
P.O. 5288
Salem, Oregon 97304

Wards

9.3 Passive Solar Water Heaters

Zomeworks
Breadbox Plans and Passive Solar Water Heaters
P.O. 712
Albuquerque, NM 87103

10. Daylighting

Solar Lighting Research
Glendale, AZ

11. Solar Electric

11.1 Wind

The American Wind Energy Association $25/yr
54468 CR 31
Bristol, IN 46507

Earthmind
4844 Hirsch
Mariposa, CA 95338

Wind Power Digest $6.00 sub. with catalog
54468 C.R. 31
Bristol, IN 46507

Real Gas and Electric
P.O. Box F
Santa Rosa, CA 95402

11.2 Photovoltaic

Energy Conversion Devices
Troy, Michigan

Solar ex
1335 Piccaro Dr.
Rockville, MD 20850

Real Gas and Electric
P.O. Box F
Santa Rosa, CA 95402

Energy Saving Systems
Price and Pine Streets
Holmes, PA 19043

Energy House
P.O. 5288
Salem, Oregon 97304

11.3 Solar Thermal

Ormat Systems Inc.
209 West Central Street
Natic, Massachusetts 01760

11.4 Solar Pump

Girardier
Establishments Pierre Mengin
Amilly, France

12. Backup Systems

12.1 Fans and Blowers

The Broan Co.
926 W. State Street
Hartford, WI 53027

Thermo-Attix
800 Industrial Rd.
San Carlos, CA 94070

Dale and Associates Inc.
1401 Cranston Rd.
Beloit, Wisc. 53511

W.W. Grainger, Inc.
5959 W. Howard
Chicago, IL 60648

The Muffin Fan
Rotran Inc.
Hasbrouck Lane
Woodstock, N.Y. 12498

The Universal Circ. Fan
Martin Ind. Div.
Sheffield, ALA. 35660

12.1 Ceiling Fan - Casablanca

Brass Bed Factory
4120 Sepulveda
Century City, CA 90067

Casablanca Fan Co.
182 S. Raymond
Pasadena, CA

Fly Fan Factory
2099 Westwood Blvd.
Westwood, CA 90025

George Town Fan Co.
P.O. Box 2128-T
Denver, CO 80201

Hunter
2500 Frisco Avenue
P.O. Box 14775
Memphis, TN 38114

Montgomery Wards

Sears

12.3 Heat Exchanger - Ventilators

Manufacturer

Aeroplast
Azay le Brule
F-79400 S
St. Maixent
France

Air Froehlich
CH-9320 Arbon
Switzerland

Aldes Industries
82 rue Feuillat
F-69008 Lyon
France

Bacho Ventilation
S-19901 Enkoeping 1
Sweden

R.W. Besant
Department of Mechanical
Engineering
University of Saskatchewan
Saskatoon, Canada S7N OWO

V.M. Christensen A/S
Klimatenknisk AFD
Febriksparken 22
DK-2600 Glostrup
Denmark

AB Svenska Flaktfabriken
Equipment Division Research
Office
Asenvagen 7
S-55 184 Jonktoping
Stockholm, Sweden

Happel
Germany (address unknown)

Harrison Radiator Division
General Motors Corporation
Lockport, New York

Mitsubishi Electric Corp.
2-3 Marunouchi
2-Chome
Chiyoda-Ku
Tokyo, Japan

Norlett
A.S. Norlett
1801 Askim, Norway

Q-Dot Corporation
151 Regal Row
Suite 220
Dallas, Texas

12.4 Vents

H.C. Products, Co.
P.O. 68
Princeville, Ill. 61559

Triangle Engineering
P.O. 7464
Houston, TX 77008

Penn Ventilator
11th & Allegheny Ave.
Philadelphia, PA 19140

Airstream Vents
3701-55 Sepviva St. at Colston Ave.
Philadelphia, PA 19137

Hart & Cooley - Registers & Grills
Allied Thermal Cap
Holland, Mich 49423

12.5 Backup Heating

A. Wood
Arizona Forest Supply
P.O. Box 188
Flagstaff, AZ 86002

All Nighter Stove Works
Falcon Marketing Int.
P.O. Box 4567
Santa Rosa, CA 95401

Hayes-Te Equip. Corp
P.O. Box 266
Unionville, CT 06085

Ashley Products Division
Martin Industries
1604 17th Ave, S.W.
P.O. Box 760
Sheffield, AL 35660

Energy Harvesters Corp
P.O. Box 19
Fitzwilliam, NH 03447

The Hearthmate
C&D Distributer Inc.
P.O. Box 766
Old Saybrook, CT 06475

France Belge Foundries
70 Pine Street
New York, NY 10005

Kristia Associates $2
P.O. 1118
Portland, ME 04104

Vermont Castings Inc. $1
Prince St.
Randolph, VT 05060

Tempwood
Mohawk Industries
P.O. Box 71-0
Adams, MA 01220

Tirolia Cook Stove
P.O. Box 330
Vacaville, CA 95688

Waterford Wood Stoves
P.O. Box 2407
Grand Central Station
New York, NY 10017

Morso
Southport Stoves
248 Tolland St.
E. Hartford, Conn. 06108

Buffalo Blower
Columbia Products
205 W. Monroe St.
Chicago, Ill. 60606

Heatilator
19118 Saunders
Mt. Pleasant, Iowa 52641

The Lorena Mud Stove $3
ATP - Volunteer in Asia
Box 4543
Stanford, CA 94305

Schrader Stoves
P.O. 2898
Eugene, Ore. 97477

Lynndale Central Heating Systems
1100 N. University Suite 139
Little Rock, Ark. 72207

Gardenway Catalog
Charlotte, VT 05445

The Majestic Co.
P.O. Box 800
Huntington, Ind. 46750

B. Active Solar

Timberline Energy Corp.
1840 Le Moyne Ave.
Syracuse, NY 13208

American Solarize
139 South Street
Murray Hill, NJ 07974

Grumman Sunstream
4175 Veterans Memorial Highway
Ronkonkoma, NY 11779

Solar Age Catalog (1977) $8.50
Bruce Anderson and Sandra Oddo
Solar Age
P.O. Box 305;Dover, NH 07801

Solaris
609 Cedar Ave
Fort Washington, MD 20022

Solaron Corp.
300 Galleria Tower
720 S. Colorado Blvd.
Denver, CO 80222

12.6 Evaporative coolers

Fedders
Edison, NJ 08817

Int'l Metal Products Division
McGraw Edison
P.O. 20188
Phoenix, AZ 85036

12.7 Water Heater Backup

Intermatic Inc. - Setback Water Heater
Plaza
Spring Grove, Ill. 60081

Rheem
7600 S. Kedzie Avenue
Chicago, Ill. 60652

Blazing Showers
P.O. 377
Point Arena, CA 95468

Waterford (see wood stoves)

Energy House
P.O. 5288
Salem, Oregon 97304

American Appliance
2341 Michigan Ave.
Santa Monica, CA

TELL THEM YOU HEARD ABOUT THEM
IN THE PASSIVE SOLAR CATALOG

13. Appliances

13.1 Refrigerators, Stoves, etc.

Amana Ref. Co.
Amana, IA 52203

Philco-Ford
Union Meeting Road
Blue Hill, PA 19422

In-Sink-Erator (I.S.R.)
4700 21st Street
Racine, Wi 53406

The Maytag Company
17065 E. Green Drive
City of Industry, CA 91744

Kitchen Aid
Hobart Mfg. Co.
1501 W. 8th Street
Los Angeles, CA 90017

13.2 Composting Toilets

A.B. Gustasvsbergs Fabriker
Flemingatan 62 B, Box 12159
10224 Stockholm 12, SWEDEN

Bio-Let
Bio-Utilities, Inc.
Box 135
Narbeth, PA 19072

Clivus Multrum USA, Inc.
14A Elliot Street
Cambridge, MA 02138

Green Earth Manufacturing
80996 Pacific Hwy, North
Creswell, OR 97426

Pacific Ecological Systems
P.O. Box 10883
Eugene, OR 97401

Enviroscope, Inc.
Box 752
Corona Del Mar, CA 92625

Humus Toilet 5
Ecos, Inc.
21 Imrie Rd.
Boston, MA 02134

13.3 Water Conservation

Flush-Gard
Utah Marine
459 South 7th St.
Bringham City, UT 84302

MICROPHOR "Two qt. Toilet"
P.O. Box 490
Willits, CA 95490

The Flushmate-N.A. Inc.
1100 Owendale, Suite E
Troy, MI 48048

Minuse Enviro-Systems, Inc.
206 N. Main Street, Suite 300
Jackson, CA 95642

14. Instruments

Taylor Instrument
Consumer Products Division
Sybron Corp.
Arden, NC 28704

Climet Instrument Co.
P.O. Box 151
Redlands, CA 92373

Turtox/Cambosco
MacMillan Science Co., Inc.
8200 South Hoyne Avenue
Chicago, IL 60620

MISCO
1825 Eastshore Highway
Berkeley, CA 94710

Weathertronics
2777 Del Monte St.
West Sacramento, CA 95691

Towner Systems
14666 Doolittle Drive
San Leandro, CA 94577

Meterology Research Inc.
P.O. Box 637
Altadena, CA 91001

Belfort Instrument Co.
1600 South Clinton St.
Baltimore, MD 21224

Bailey Instruments
P.O. Box 5068
Saddle Brook, NJ 07662

Barnes Engineering Co.
30 Commerce Road
Stamford, CT 06904

Entropy Limited $1
5735 Arapahoe Avenue
Boulder, CO 80303

Eppley Labratory Inc.
12 Sheffield Ave.
Newport, RI 02840

Rho Sigma
15150 Raymer St.
Van Nuys, CA 91405

A.M. Leonard and Sons
Piqua, OH 45355

Edmund Scientific
7780 Edscorp Bldg.
Barrington, NJ 08007

E.G. Geiger
P.O. Box 285
Harleysville, PA 19438

NASCO
1524 Princeton Avenue
Modesto, CA 95392

Science Associates, Inc.
P.O. Box 230
Princeton, NJ 08540

Hollis Observatory
One Pine Street
Nashua, NH 03060

Texas Electronics Inc.
P.O. Box 7225
Dallas, TX 75209

Weathermeasure
Box 41257
Sacramento, CA 95841

Science Associates Inc.
230 Nassau Street
P.O. Box 230
Princeton, NJ 08540

Dwyer Instruments Inc.
Michigan City, IN 46360

R.E.I.
P.O. Box C 88125
Seattle, WA 98188

Sierra Envir. Products
1825 Eastshore Highway
Berkeley, CA 94710

DATALOGGER
Norman Saunders
15 Ellis
Weston, MA 02193

15. The Good Life

Intermediate Technology
556 Santa Cruz Avenue
Menlo Park, CA 94025

Intermediate Tech. Devel.
9 King Street
London
ENGLAND

The National Center for Appropriate Tech.
P.O. Box 3838
Butte, Montana 59701

App. Tech. Sourcebook $4
Volunteers in Asia
Box 4543
Stanford, CA 94503

Minimum Cost Housing Group
School of Architecture,
McGill University
3480 University St.,
Montreal, Quebec H3A 2A7
Canada

Soft Tech (1978) $5
Jay Baldwin and Stewart Brand
Co Ev Quartery
Box 428
Sausalito, CA

The Complete Biogas Handbook
At Home Everywhere No. 5
Rt. 2 Box 295A
Aurora, Oregon 97002

Jojoba Oil & Seed Co.
1407 S. Date Street
Mesa, AZ 85202

The Old House Journal $12 yr.
Dept. 7
199 Berkeley Place
Brooklyn, N.Y. 11217

The New Farm $10
33 East Minor
Emmaus, PA 18049

Bicycling $6.99 yr.
33 East Minor
Emmaus, PA 18049

Wilderness Camping $7.50 yr.
1597 Union Street
Schneutady, N,Y. 12309

The Country Journal $12 yr.
139 Main Street.
Brattleboro, VT 05301

Bikes

Bell Helmets
15301 Shoemaker Avenue
Norwalk, CA 90650

Avocet
P.O. Box 7615
Menlo Park, CA 94025

Blackburn Racks
1080 Florence Way
Campbell, CA 95008

Bicycle Lighting Systems $1 info.
718 N Vermont St.
Arlington, VA 22203

Bellwether
1161 Mission St.
San Francisco, CA 94103

Sun Tour USA
60 Chapin Road
Pine Brook, NJ 07058

Bickerton Cycle Ltd.
1314 N.W. Glisan Street
Portland, OR 97209

Running

Runners World
Box 366
Mountain View, CA 94042

Nike Shoes
8285 S.W. Nimbus
Suite 115
Beaverton, Ore. 97005

Cars

The Honda Civic CVCC
Excellent design and service

Toyota Corolla
Excellent reliability

Volvo
The lowest pollution

Tools

Woodcraft Supply $1
313 Montvale Ave.
Woburn, Mass. 01801

16. Solar Advocates and Associations

16.1 Associations

International Solar Energy Society, American Section
American Technological University
P.O. Box 1416
Kileen, Texas 76541 $25 yr.

EXISTING CHAPTERS

ALABAMA
Solar Energy Association
Dr. G.R. Guinn
c/o University of Alabama
in Huntsville
Johnson Environmental
& Energy Studies
P.O. Box 1247
Huntsville, AL 35807

ARIZONA
Solar Energy Association
Mr. Russell Clark
Secretary/Treasurer
P.O. Box 7
Flagstaff, AZ 86001

COLORADO
Solar Energy Association
Mr. Paul Shippee, President
c/o Solar Energy Program
University of Colorado
1100 14th Street
Denver, CO 80202

EASTERN NEW YORK
Solar Energy Society
P.O. Box 5181
Albany, NY 12205
(518) 457-4861

GEORGIA
Solar Energy Association
Mr. Des Yawn
Campus Box 32743
Georgia Institute
of Technology
Atlanta GA 30332

METROPOLITAN NEW YORK
Solar Energy Association
Mr. Joseph Cuba
United Engineering Center
345 East 47th Street
New York, NY 10017

MICHIGAN
Solar Energy Association
Mr. Edward J. Kelly, Jr.
201 East Liberty Street
Ann Arbor, MI 48104

MID ATLANTIC
Solar Energy Association
Mr. Robert T. Bennett
Department of Architecture
Graduate School of Fine Arts
University of Pennsylvania
Philadelphia, PA 19104

NEW ENGLAND
Solar Energy Association
Mr. John T. Schebly, Jr.
P.O. Box 541
Brattleboro, VT 05301
(802) 254-2386

NEW MEXICO
Solar Energy Association
Ms. Barbara Francis
P.O. Box 2004
Santa Fe, NM 87501

NORTHERN CALIFORNIA
Solar Energy Association
Dr. Donald W. Aitken
Director
Center For Solar Energy
Applications
San Jose State University
San Jose, CA 95192

NORTHERN ILLINOIS
Solar Energy Association
Mr. J.A. Hartley, Secretary
P.O. Box 352
Argonne National Laboratory
Argonne, IL 60439

OHIO
Solar Energy Association
Mr. Joseph J. Barbish
12931 West Geauga TRAIL
Chesterland, OH 44026
(216) 729-9350

SOUTHERN CALIFORNIA
Solar Energy Association
Mr. John T. Brand
Chairperson
202 C Street 11B
San Diego, CA 92101

TEXAS
Solar Energy Society
Mr. Peter E. Jenkins
Texas A&M University
Mechanical Engineering
Department
College Station, TX 77843

VIRGINIA
Solar Energy Association
Mr. John G. Lewis, Jr.
Chairperson
P.O. Box 12442
Richmond, VA 23241

ORGANIZING CHAPTERS

British Columbia
Canada
Colorado
Hawaii

Illinois
South-Central Illinois
Iowa
Kansas
Louisiana
Michigan
Mexico
Minnesota
Nebraska
Nevada
New Jersey
North Carolina
Oklahoma
Oregon
Puerto Rico
South Carolina
Tennessee
Metropolitan Washington. D.C.
Missouri/Arkansas

16.2 Lobbies

Solar Lobby, Rm. 1100
1028 Connecticut, D.C. 20036

Sunrae $15 yr.
P.O. Box 915
Goleta, CA 93017

Calif. Solar Action Network
216 West 8th
Davis, CA 95616
(Just starting, help by sending some money)

The Passive Solar Institute
P.O. 722
Davis, CA 95616

17. Information

17.1 Magazines With Passive Solar Info.

Organic Gardening $9 yr.
33 East Minor Street
Emmaus, PA 18049

Solar Age $20 yr.
P.O. Box 4934
Manchester, NH 03108

Mother Earth News $12 yr.
P.O. Box 70
Hendersonville, N.C. 28739

CoEvolution Quarterly $12 yr.
Box 428
Sausalito, CA 94965

Alternative Sources of Energy $10 yr.
Rt. 2
Milaca, Minn. 56353

Hudson Home Guide
289 So. San Antonio Rd.
Los Altos, CA 94022

Energy and Alternatives $4 yr.
P.O. Box 681
Salem, Ore. 97308

Rain
2270 N.W. Iring
Portland, Ore. 97210

Bulletin-NMSEA $10 yr. (membership)
P.O. Box 2004
Sante Fe, New Mexico 87501

The Solar Greenhouse Digest $ 7
P.O. 3218
Kingman, AZ 86401

Sunset $9 yr.
Lane Publishing Co.
Menlo Park, CA 94025

New Energy
8383 Willshire Blvd.
Beverly Hills, CA 96211

TELL THEM YOU HEARD ABOUT THEM IN THE PASSIVE SOLAR CATALOG

SAVE THE WHALES!

In the past fifty years alone 2 million whales have been killed. A highly mechanized commercial hunt has driven one species after another to the brink of extinction. The right whale, the bowhead, the blue and the humpback are now all depleted to 10% or less of their original populations and show little if any sign of recovery, despite the presence of international protection for more than a decade (since 1935 in the case of the right and bowhead, since 1966 in the case of the blue and the humpback). According to many experts, the blue whale - the largest animal ever to live on this planet - has been so decimated that its extinction is inevitable. You can help. Support the efforts of:

The Greenpeace Foundation, 240 Fort Mason, San Francisco, CA 94123 (415) 474-6767

Popular Science $9.94 yr.
Popular Science Subscription Dept.
P.O. Box 2871
Boulder, Colo. 80302

Popular Mechanics $7.97 yr.
P.O. Box 10064
Des Moines, Iowa 50350

The Family Handyman $8.95 yr.
P.O. 2897
Boulder, Colo. 80322ll

Seriatim $9 yr.
122 Carmel
El Cerrito, CA 94530

17.2 Solar Info Services & Books

John Peck
Tech. App. Center
2500 Central Ave. SE
Alb., New Mexico 87106

Solar Energy Info Service
P.O. 204
San Mateo, CA 94401

Cornerstones
54 Cumberland Street
Brunswick, ME 04001

National Solar Energy
Education Campaign (NSEEC)
10762 Tucker Street
Beltsville, Maryland 20725
Catalog $4.50

Owner-Builder Publications
Box 817
North Fork, CA 93643

Energy House, Catalog $3.00
P.O. 5288
Salem, Oregon 97304
A number of books &
hundreds of products

Rodale Press
Organic Park
Emmaus, PA 18049
Solar Usage Now
Box 306
Bascorn, Ohio 44809

Whole Earth Truck Store
558 Santa Cruz Ave.
Menlo Park, CA 94025

Clearinghouse for Earth Covered Buildings
School of Architecture and Environmental Design
University of Texas
Arlington, Texas 76019

Garden Way Publishing
Charlotte, Vermont 05445

Sol Pub
Box 2351
Gaithersburg, MD 20760

International Project for Soft Energy Paths
124 Spear Street
San Francisco, CA 94105

National Solar Heating And Cooling Information Center
P.O. BOX 1607
ROCKVILLE, MD 20850

National Technical Information Service
5285 Port Royal Rd.
Springfield, VA 22151

Ashrae, 345 East 47th ST. NY, NY 10017

Handbook of Fundamentals (1977) $42
Applications Handbook (1978) $42
Systems Handbook (1976) $42
Equipment Handbook (revised June 1979) $42

Energy Marketplace
Conservation Consumer
19143 15 Ave. N.W.
Seattle, WA 98177

17.3 Product Info.

Solar Products Spec. Guide $120
Solar Age Magazine
Church Hill
Harrisville, N.H. 03450

Standard & Poor's Register
345 Hudson Street
New York, N.Y. 10014

Thomas Register
One Penn Plaza
New York, N.Y. 10001

Sweets
McGraw Hill Info. Systems
1221 Ave. of the Americas
N.Y., N.Y. 10020

McRaes Blue Book
Hinsdale, ILL 60521

18. Simulation

DATA
P.O. Box 503
Davis, CA 95616

T.E.A.-Net
Church Hill
Harrisville, N.H. 03450

Gard, Inc.
7449 North Natchez Ave.
Niles, IL 60648

Berkeley Solar Group
3026 Shattuck
Berkeley, CA 94705

Phil Niles
Rt. 2 Box 714 R
Arroyo Grande, CA

Princeton Energy Group
729 Alexander Road
Princeton, NJ 08540

VILLAGE HOMES SOLAR HOUSE DESIGNS

This is the only book about a solar subdivision and, as such, it provides insight into what is to come in housing developments. Solar designs frequently are so expensive that their use is limited to high-budget homes. But in VILLAGE HOMES, the conservation minded homeowner has an opportunity to learn about energy-efficient, solar-heated and-cooled housing designed for middle-income families. The book provides extensive text, diagrams, photographs, and illustrations of 43 homes in Village Homes, a solar subdivision in Davis, California.

What makes VILLAGE HOMES so unique is that it is concerned with innovative house and community designs as they apply to a *working* community. All the houses described have actually been built and are now being lived in. Both successes and flaws of the systems are presented by their designers and residents.

The first part of the book focuses on the planning principles behind Village Homes and emphasizes land use, landscaping, and community design. The second part deals with the homes and discusses the various water heaters—passive and thermosiphon, and active—used in the houses, as well as the passive and active solar design theories and practices incorporated in the houses. Highlighted are design features such as clerestories, water walls, skylights, suncatchers, greenhouses, and active water systems. Complete with descriptive text and plenty of detailed illustrations and photos, the book is a must for anyone interested in ecological, contemporary housing.

by David Bainbridge, Judy Corbett, and John Hofacre

Paperback $6.95

From: Rodale Press May 1979
Organic Park
Emmaus, PA 18049

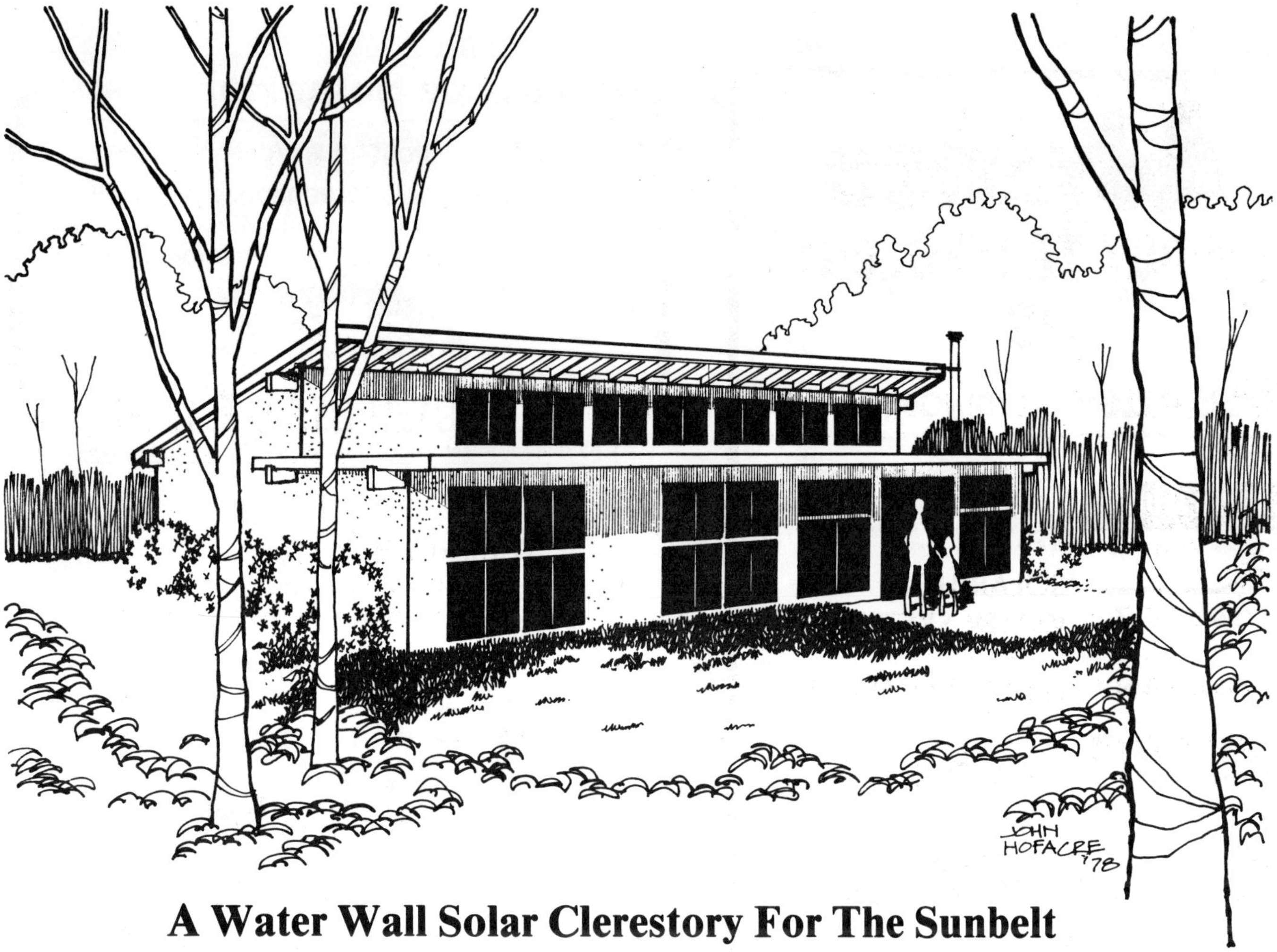

A Water Wall Solar Clerestory For The Sunbelt

To help builders and developers get started in Passive Solar building, the Passive Solar Institute commissioned John Hofacre to design a simple, 3 bedroom, 2 bath water wall house for the sunbelt. Drawing upon his experience from more than 40 passive solar houses he has designed and seen built, John developed a very pleasing solar clerestory.

After the houseplans and details were finalized, the Passive Solar Institute engaged Davis Alternative Technology to evaluate the performance of this passive solar home in 10 cities in the areas of fastest growth in the U.S. The results were impressive and show very clearly what can be done without "radical" or expensive designs.

Performance of the Solar Clerestory Water Wall House

Area	Solar Heating	Natural Cooling
Los Angeles, CA	100%	100%
Sacramento, CA	94%	100%
Medford, OR	72%	100%
Phoenix, AZ	100%	50%
Denver, CO	82%	100%
Ft. Worth, TX	100%	30%
Little Rock, AK	89%	37%
Knoxville, TN	75%	100%
Raleigh, NC	99%	43%
Miami, FL	100%	57%

Even better performance and greater savings could be achieved in a more sophisticated passive design. The flat roof on the south of the clerestory was included so that a more sophisticated cooling system could be installed without complete revision of the plan. A Skytherm" system, "Energy Roof", flooded roof, or sod roof could all provide additional cooling in the hotter areas. The added mass of these systems would require revisions to the framing and ceiling and should be worked out with the system designer or a structural engineer.

Other plans will be developed in 1979 for colder climates. Send a self-addressed stamped envelope for details (after March1, 1979).

Plans for the Solar Clerestory and 1/8" scale printed black on white paper can be ordered from the Passive Solar Institute. Three copies will be included. Ask for Plan 78-1 and enclose a check or money order for $25.00. Expect delivery in two to four weeks.

Solar House Plans, Dept. PSC
The Passive Solar Institute
P.O. Box 722
Davis, CA 95616

TELL THEM YOU HEARD ABOUT THEM IN THE PASSIVE SOLAR CATALOG

TELL THEM YOU HEARD ABOUT THEM IN THE PASSIVE SOLAR CATALOG

Standard Items

Standard BubbleLite Plastic Skylights

Calif. State Fire Marshal Listing No. 4350-035:4

Special Order Items

O'Keeffe's, Inc.
75 Williams Avenue
San Francisco 94124
(415) 822-4222

Also available upon request: roof hatches, fire and smoke vents, access ladders and reglets.

Send one dollar to the above address for a descriptive brochure.

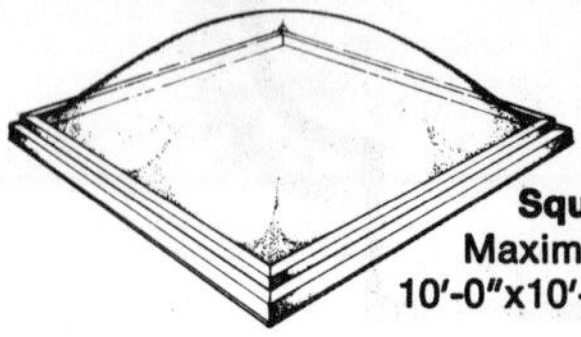

Square Dome
Maximum Size—
10'-0"x10'-0"

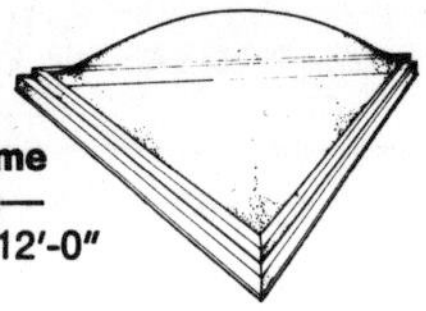

Triangular Dome
Maximum Size—
12'-0"x12'-0"x12'-0"

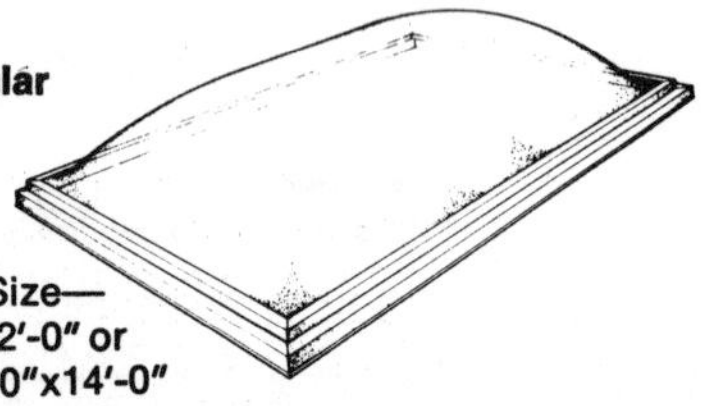

Rectangular Dome

Maximum Size—
10'-0"x12'-0" or
8'-0"x14'-0"

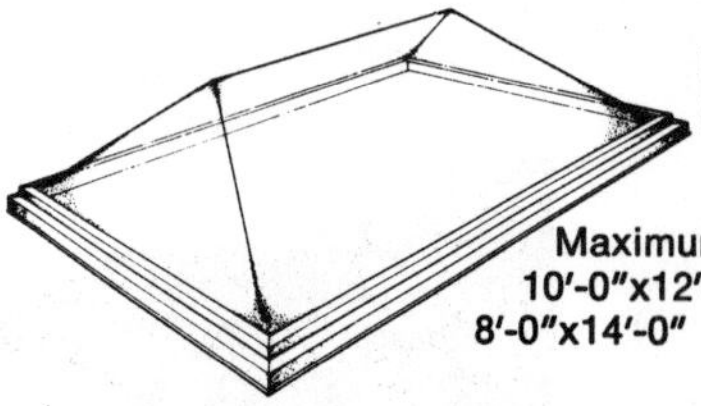

Double Hip Type

Maximum Size—
10'-0"x12'-0" or
8'-0"x14'-0"

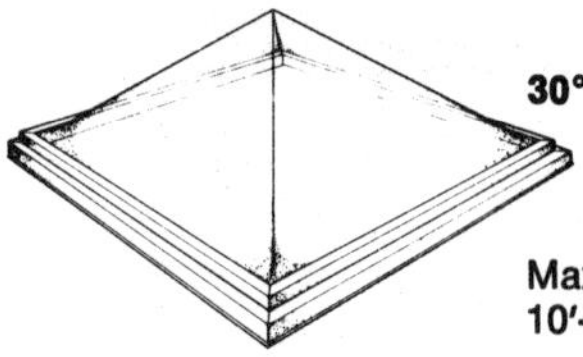

30° Pyramid Type

Maximum Size—
10'-0"x10'-0"

45° Unit Available Up To 8'-0"x8'-0"

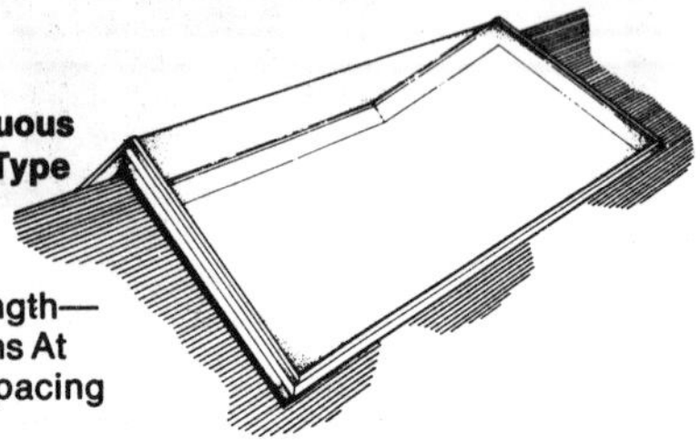

Continuous Ridge Type

Any length—
Mullions At
6'-0" Spacing

O'Keeffe's **"BubbleLite"** plastic skylights provide a practical, economical means to utilize free natural daylight, saving thousands of kilowatt hours of energy annually on electric bills. They can flood a dismal kitchen with light; illuminate a dark hallway; brighten a workshop and let you see what you are doing. An O'Keeffe **"BubbleLite Skywindow"** can also provide a room with ventilation, as it fills it with light.

Many sizes are produced in quantity and are normally available from stock. However, larger sizes, and any standard size not in stock, can be produced in an average maximum of two weeks.

In some instances you may not find the exact size you need from our list of standard sizes. In that case, any size can be manufactured to your exact requirements, with a small additional charge for custom tooling.

Standard lenses are Optically Clear (92% light and 90% solar energy transmittance); White Translucent (56% light and 52% solar energy transmittance); and Optically Clear Bronze (27% light and 35% solar energy transmittance).

Standard is aluminum color, mill finish as extruded. For additional protection against corrosion, as well as enhanced appearance, we recommend a thermosetting enamel color coating. Standard colors available are: light bronze; medium bronze; dark bronze; matte black; or matte white.

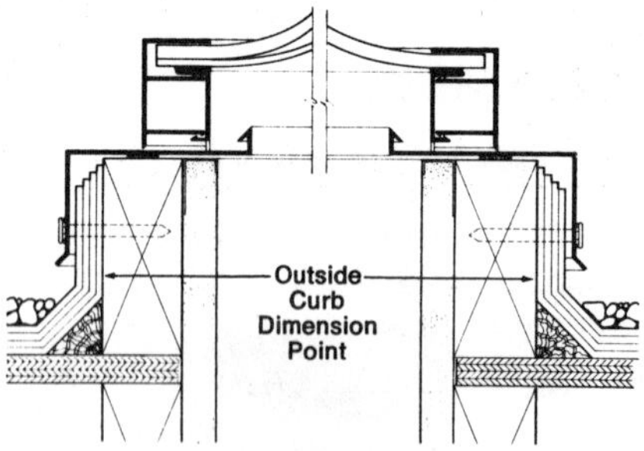

Model No. 3
Standard Curb Type
(For Units Under 4'-0"x4'-0")
Available with Single or Double Acrylic Lens
• Square Domes • Rectangular Domes
• Pyramids

Calif. State Fire Marshal Listing No. 4350-035:4

Scale: 3" = 1'-0"
See Full Size Details

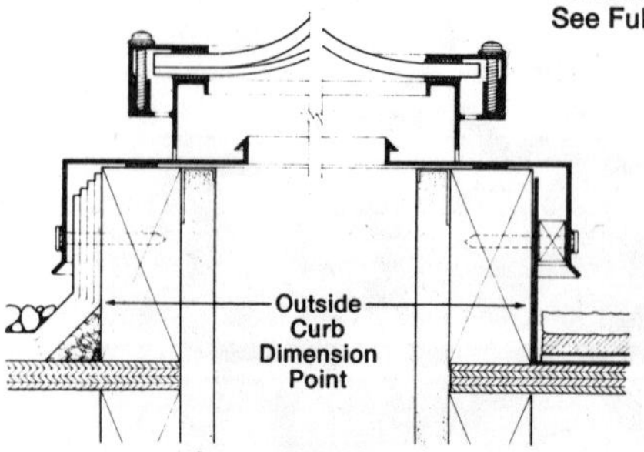

Model No. 6
Standard Curb Type
(For Units Over 4'-0"x4'-0")
Available with Single or Double Acrylic Lens
• Square Domes • Double-Hip Units
• Pyramids • Continuous Ridge Units

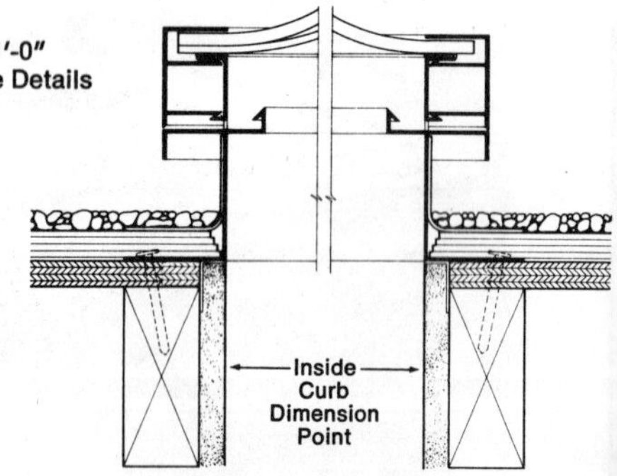

Model No. 11
Integral Curb Type
(For Units Under 4'-0"x4'-0")
Available with Single or Double Acrylic Lens
• Square Domes • Rectangular Domes
• Pyramids

PASSIVE SOLAR-VARIANT HOMES BY SAVELL

Passive and Active Solar Systems Defined/Compared

A PASSIVE SOLAR HEATING AND COOLING SYSTEM IS DEFINED AS ONE IN WHICH ENERGY FLOW IS ENTIRELY BY NATURAL MEANS, maximizing the use of natural convection, conduction and radiant energy such as that found in the Egyptian pyramids and early American Pueblos.

In contrast, the majority of systems being designed today are the active systems which are usually an unsightly collection of solar panels, pumps, fans, ducts, insulation, massive heat storage areas and controls which are combined to move solar energy from collection points to a storage area of water, rocks or salts, later to be released into the home through a system of ducts, fans etc.

COSTS: ACTIVE vs. PASSIVE

Potentially, the active systems require more maintenance and upkeep than passive systems. On the other hand massive walled homes, such as the Adobe and underground structures, are expensive to build and impractical in many parts of the world due to seismic conditions and severity of cold weather. The initial first cost of either system increases the basic price of an average home and is seldom cost effective over the immediate future. Affordable housing is therefore beyond the financial capability of many.

THE SAVELL SYSTEM

The Savell system utilizes the passive approach to thermal control by creating a massive structure of concrete walls and floors and applying insulation to the exterior concrete wall panels. Our storage medium (concrete) therefore becomes functional and avoids the high first cost associated with active and passive systems. Improved energy efficiency IMMEDIATELY reduces home operating costs and makes these homes MORE COST EFFECTIVE THAN EVEN CONVENTIONAL CONSTRUCTION METHODS!! (See graph)

Exterior walls are finished with stucco or other decorative treatment. Interior concrete walls are finished with textured paint, paneling, paper, etc. This encapsulated structure thus becomes a massive heat sink that overwhelms it's environment and causes the structure to function in equilibrium (within very narrow limits) with the earth. Before the heat of the day can effect a change, the day is over! The need for additional energy (during any season) to "fine tune" the interior environment for maximum comfort is minimal! In this PATENTED method of construction, *what you see is what you get!!* There are no unsightly displays of solar collection devices!

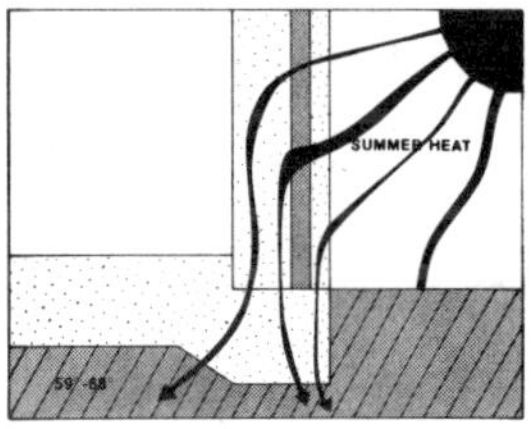

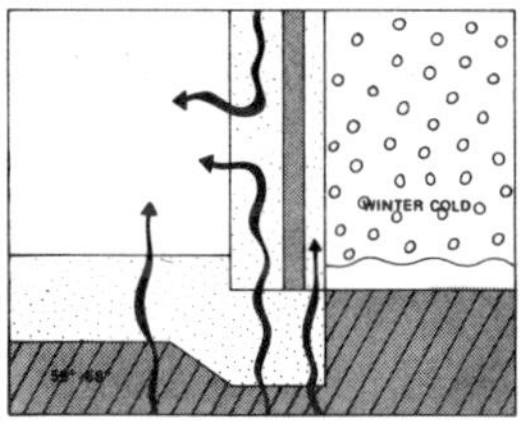

THE CONSTANT TEMPERATURE LINE. The earth has what is called a Constant Temperature Line several feet beneath the surface, below which temperature remains steady, summer and winter. By creating a massive structure, mating it to the earth by the heavy slab, and then by *encapsulating it in an insulating cocoon,* we have effectively raised the Constant Temperature Line *up* through the conductive walls, making the house more a part of the earth than of the atmosphere. During summer, little or no cooling is required; during winter, minimal heat will suffice.

The illustration above demonstrates this phenomenon.

WINTER HEATING. During the winter the concrete walls of the house are charged with energy, either from solar gains through windows or small space heaters, as well as from body heat and the heat from cooking or lighting systems. Since the walls will store heat and radiate it back into the house as night temperatures fall, little auxillary heat is required.

SUMMER COOLING. Just as water seeks its own level so does heat flow to cold. The protective cocoon of insulation which retains winter heating also keeps summer heat out. However, despite the best insulation, infiltration of summer heat occurs through the walls, windows, doors and roof. In a Savell home this excess heat flows into conductive walls and is subsequently dissipated through the floors into the cooler earth: the reverse of winter operation.

PASSIVE SOLAR VARIANT HOMES BY SAVELL
575 Birch Court, Suite A, Colton, California USA, 92324, Tel. (714) 825-4615

TELL THEM YOU HEARD ABOUT THEM IN THE PASSIVE SOLAR CATALO

SOLAR GREENHOUSE DIGEST

EDITOR & PUBLISHER
TWILA de VRIES

(602) 757-8255

P.O. BOX 3218
KINGMAN, AZ 86401 U.S.A.

A NEW SOLAR ENERGY JOURNAL

SOLAR GREENHOUSES

- *new approaches to heat and food production*
- *for families or large growers*

GARDEN AT WAIST LEVEL ALL YEAR!
ADAPTABLE TO DIFFERENT CLIMATES

- *using little or no energy*
- *attached or unattached*

BUYING, BUILDING, GROWING OR HOBBY SOLAR GREENHOUSES ARE FOR YOU!

ALTERNATE ENERGY—BE INDEPENDENT! LEARN ABOUT SOLAR, WIND, HYDROGEN, GASAHOL, AND OTHER ALTERNATIVES

JOIN THE GREENHOUSE REVOLUTION!

SUBSCRIBE TO THIS INTERNATIONAL MAGAZINE NOW!

SOLAR GREENHOUSE DIGEST, P.O. BOX 3218, KINGMAN, ARIZONA 86401 - (602) 757-8255

Clip & Mail

Subscription rates: 1 year $7.00 ________ 2 years $12.00________

Add $1.00 for Canada and $2.00 for foreign countries, $5.00 air mail per year. U.S. Money Only.

Please send an introductory copy of SOLAR GREENHOUSE DIGEST. Enclosed is $1.00 to cover postage and handling.

Please Charge My: ☐ *MASTER CHARGE* ☐ *VISA* *Card No.* ____________________

Expiration Date ____________

NAME ____

ADDRESS ____________________ CITY ____________ STATE ______ ZIP ________

ORGANIZATION ______________________________ PHONE ______________

Please direct all requests to SOLAR GREENHOUSE DIGEST, INC., P.O. BOX 3218, Kingman, Arizona 86401 (USA)

PROMOTE SOLAR ENERGY WITH. . .

★ **BUMPER STICKERS. . .** SOLAR ENERGY **.50** ☐

DECALS. . ..25 ☐ **BUTTONS. . ..50** ☐ **PATCHES. . .$2.00** ☐

SOLAR HATS (ADJUSTABLE) AVAIL. IN YELLOW, GREEN, COLUMBIA BLUE, NAVY . . . . **$5.00** ☐

T-SHIRTS

#1	SOLAR ENERGY	AVAILABLE IN CREAM, BEIGE, YELLOW & LIGHT BLUE	MENS $6.00	☐
			WOMENS $7.00	☐
			CHILDRENS $5.00	☐
#2	SOLAR FUTURE	AVAILABLE IN CREAM, BEIGE, YELLOW & LIGHT BLUE	MENS $6.00	☐
			WOMENS $7.00	☐
#3	THE SAFE NUCLEAR REACTOR	NAVY BLUE ONLY	MENS $6.00	☐
			WOMENS $7.00	☐
#4		BEIGE ONLY	MENS $6.00	☐
			WOMENS $7.00	☐

★ **PASSIVE SOLAR CATALOGUE. . . . .$5.00** ☐

★ **HOME ENERGY CONSERVATION KIT** ☐

$24.00 INCLUDES CUSTOM COMPUTER ANALYSIS
$ 6.00 EXTRA FOR SOLAR ANALYSIS
$ 6.00 FOR SOLAR POOL HEATING ANALYSIS

STAINED GLASS WINDOW HANGERS

(7½″ DIAMETER)

17 PC. . . . .$20.00 ☐ 20 PC. . . . .$24.00 ☐

($4.00 SHIPPING CHG. INCLUDES PRIORITY MAILING INSURANCE & HANDLING)

PLEASE SEND ME THE ITEMS I HAVE INDICATED

ENCLOSED IS CASH ☐ CHECK ☐ MONEY ORDER ☐ FOR FULL AMOUNT PLUS $1.30 HANDLING ON HATS & T SHIRTS (.50 ON ALL ELSE). CALIF. RES. ADD 6% SALES TAX.

NAME ____________________
ADDRESS ____________________
CITY ____________________
STATE ____________ ZIP ________

SEND TO
SOLAR FUTURE
P.O. BOX 19177
SACRAMENTO, CA 95819

WHOLESALE PRICES AVAILABLE

2: Consultants

The listing of passive consultants in this section in no way constitutes a recommendation or warranty of their services. Like all professional consultants they vary widely in terms of ability, knowledge, and skills. Before you hire a passive consultant evaluate your skills carefully and you may discover you don't need one. If you feel inadequate for the task of designing your own house or system check with previous clients before signing a contract. Did his design work? Was it buildable? Was it on time? And consider these words from Vitruvius concerning cost:

"(in the Greek city of Ephesus) When an architect accepts the charge of a public work, he has to promise what the cost of it will be. His estimate is handed to the magistrate, and his property is pledged as security until the work is done. When it is finished, if the outlay agrees with his statement, he is complimented by decrees and marks of honor. If no more than a fourth has to be added to his estimate, it is furnished by the treasury and no penalty is inflicted. But when more than one fourth has to be spent in addition to the work, the money required to finish it is taken from his property.

Would to God that this were also a law for the Roman people, not merely for public, but also for private buildings . . . architects themselves could be forced, by fear of the penalty, to be more careful in calculating and stating the limit of expense."

If you want information from these generally good-hearted, impoverished folks, send them a self-addressed, stamped envelope or postage and at least $10. That'll get you at least one half hour of review time.

Make your question as clear as possible. I've tried to answer a number of letters of this type over the years and some are impossible, such as:

"I'm going to build a house next year (maybe). I plan to do a solar house using phase change materials for storage. What do you think of the plan below:"

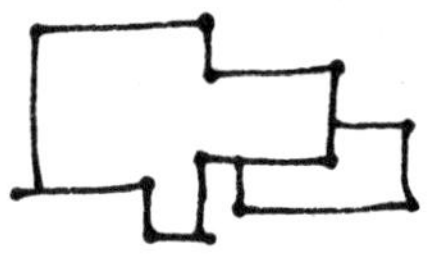

My question to them: "Is this a system diagram or a house plan? Will it be in the sun? How do you expect to use the phase change? etc.

Do your homework and ask a responsible question, and expect to pay for a detailed answer. It will often be very worthwhile. A check of your calculations of performance and detailing may cost $50-100 or more, but it may save you several hundred dollars and uncalcuable grief.

Valerie Anderson
2404 Bucklebury
Davis, CA 95616

Blair Abee
269 Cardinal Rd.
Mill Valley, CA 94941

ARC Sun Design Associates
5907 Garlough Drive
San Jose, CA 95123

Advance Cooler Mfg. Co.
P.O. Box 387
Clifton Park, N.Y. 12065

Architects Group
310 Steele St.
Denver, Co. 80206

Approvecho Institute
359 Polk Street
Eugene, Oregon

Barkman Engineering
Cienaga Street
Santa Fe, New Mexico 87501

David Bainbridge
2446 Bucklebury
Davis, CA 95616

Thomas Bligh
Dept. of Civil & Min. Eng.
University of Minnesota
Minneapolis, Minnesota

Mr. Mike Boylson
Alfalfa Hill-Star Route
Manton, CA 96059

Burt-Hill, Kosar & Rittelmann
400 Morgan Center
Butler, PA 16001

Justin A Bereny
P.O. Box 204
San Mateo, CA 94401

Steve Baer
P.O. Box 712
Albuquerque, NM 87013

Berkeley Solar Group
3026 Shattuck
Berkeley, CA 94710

Center for Maximum Potential
Building Systems
8604 Webberville Rd.
Austin, TX 78724

Mr. J.I. Clairk, AIA, AIP
c/o Jica Arich. & Assoc.
1826-A South Elena Avenue
Redondo Beach, CA 90277

George Cody Architect
212 High Street
Palo Alto, CA 94301

Douglas Coonley
CHI Housing, Inc.
Box 566
Hanover, NH 01755

Mr. Bruce Corson
1813-A Francisco Street
Berkeley, CA 94703

Mr. Richard Cushing
5907 Garlough Drive
San Jose, CA 95123

Polly Cooper
2065 McCollum
San Luis Obispo, CA 93401

Clearinghouse for Earth Covered Buildings
School of Architecture
University of Texas
Arlington, TX 76019

Mr. Bruce Davis
408 Don Miguel
Santa Fe, NM 87501

Mr. David R. Dolen
624 El Terino Avenue
Modesto, CA 95350

Design Alternatives Inc.
1312 18th Street, N.W.
Washington, D.C. 20036

Ianto Evans
359 Polk Street
Eugene, Oregon 97402

e³ Environmental Architects
David Wright AIA
P.O. Box 49
The Sea Ranch, CA 95497

Eugene Eccli
Design Alternatives Inc.
1312 18th Street N.W.
Washington, D.C. 20036

Energy Management Consultants, Inc.
6380 Wilshire Blvd., Suite 1407
Los Angeles, CA 90048
c/o Douglas Stenhouse, AIA, AIP

Energy Management Consultants, Inc.
4055 South Spencer, Suite 226
Las Vegas, Nevada 89109
c/o John Sausen, P.E.

Energy Managment Consultants
666 State Street
San Diego, CA 92101
c/o Stanley Keniston, AIA

Energy Managment Consultants
2150 Shattuck Ave.
Berkeley, CA 94704
c/o Harvey J. Bryan AIA

Earth/Life System Design
Architecture & Urban Design
1637 Seal Way
Seal Beach, CA 90740

EKOSE'A
3375 Clay Street
San Francisco, CA 94118

Lauren Elliott
10429 Moonshine Rd.
Sebastapol, CA

Farallones Institute
15290 Coleman Valley Road
Occidental, CA 95464

Mr. Jacob Feldman
San Luis Obispo Solar Group
Calif. Poly. State Univ.
San Luis Obispo, CA 93407

Paul Fellers
800 South Third St.
Bozeman, MT 59715

Mr. Drew Gillett
319 Holbrook Road
Bedford, NH 03102

Green Mountain Homes
Royalton, Vermont 05068

Mr. John Hammond
Living Systems
Route 1, Box 170
Winters, CA 95694

Ken Haggard
2065 McCollum
San Luis Obispo, CA 93401

Mr. Robert Hastings
Architect
11571 Links Drive
Reston, VA 22090

Mr. John Hofacre
2437 Elendil Lane
Davis, CA 95616

Housing Research
3707 Jeanne Mance
Montreal, Canada H2X2K4

Marshall Hunt
2312 Elendil
Davis, CA 95616

The Habitat Center
573 Mission St.
San Francisco, CA 94105

Jan Hamrin, Director
Energy Extension Service
University Extension
Davis, CA 95616

Charles Huddleston
805 Seaview Drive
El Cerrito, CA 94530

Harold Hay
Skytherm Processes
2424 Wilshire Blvd.
Los Angeles, CA 90057

Interactive Resources, Inc.
117 Park Place
Point Richmond, CA 94801

Janice Jacobson
216 West 8th St.
Davis, CA 95616

M. Dean Jones
Architecture - Community Planning
45 Sycamore Ave.
Mill Valley, CA 94941

Kalwall
P.O. Box 237
Manchester, NH 03105

Lawrence Kasparowitz
Sierra Nevada Design Studios
P.O. Box 768
Penngrove, CA 94951

Dave Koenigshofer
Koenigshofer Engineers
1842 8th Avenue
Sacramento, CA 95818

Bill Kopper
The House Doctor
1313 Colgate Drive
Davis, CA 95616

Stanley Keniston, AIA
Architecture/Energy Systems
666 State Street
San Diego, CA 92101

Ken Kern
Box 817
North Fork, CA 93643

Paul Lutton
406 Main St.
Vacaville, CA. 95688
and
50 St. James Place
Piedmont, CA. 94611

Dennis Long
Rt. 1 Box 158
Woodland, CA. 95695

Willis & Lillian Leenhouts, Architects
3332 N. Dousman Street
Milwaukee, WS 53212

Mr. Don Loomis
1900 Mar West
Tiburon, CA 94920

Mr. William Lumpkins
535 Camino del Monte Sol
Santa Fe, NM 85701

Ted Lucas
10371 Stone River Court
Fountain Valley, CA 92708

Mr. Bruce Maeda
Davis Alternative Tech. Associates (DATA)
P.O. Box 503
Davis, CA 95616

Bruce Melzer
223 University Avenue
Davis, CA 95616

Mr. Gary Millhollen
5838 Robertson Avenue
Carmichael, CA 95608

Professor Murray Milne
Architecture & Urban Planning
University of California
Los Angeles, CA 90024

Mr. William Mingenbach
P.O. Box 49
Taos, NM 87571

Edward Mazria and Associates
P.O. 4883
Albuquerque, NM 8707

Frank Moreland
School of Architecture
University of Texas
Arlington, Texas 76019

Ron McClure
Modulor Structures Ltd.
Box 628
Tijeras, NM 87059

Tim Maloney
One Design Inc.
Mt. Falls Route
Winchester, VA 22601

Harry Miller
808 Cerrito Ct.
Albany, CA 94706

Minimum Cost Housing Group
School of Achitecture, McGill Univ.
3480 University St.
Montreal, Quebec H3A 2A7

Horace McCracken
329 W. Carlos
Alturas, CA 96101

Don Metz Earth Tech
Pinnacle Rd.
Lyme, N.H. 03768

National Center for Appro. Technology
P.O. Box 3838
Butte, MT 59701

NET ENERGY
630 - 9th Street
Arcata, CA 95521

New Alchemy Institute
P.O. Box 47
Woods Hole, MA 02543

New Mexico Solar Ene. Assn.
P.O. Box 2004
Santa Fe, NM 87501

Natural Heating Systems
2417 Front Street
West Sacramento, CA 95691

Wayne & Susan Nichols
RR #3, P.O. B 81D
Santa Fe, NM 87501

Mr. Earl Norcross
Savell Systems Intl.
P.O. Box 6305
Santa Barbara, CA 93111

Northern California Solar Energy Assn.
P.O. Box 1056
Mountain View, CA 94042

Lynn Nelson
573 Mission St.
San Francisco, CA 94105
415-543-1294

Phil Niles
Route 2 Box 714R
Arroyo Grande, CA

The Owner-Builder Center
1824 A 4th St.
Berkeley, CA 94510

Pacific Sun, Inc.
540 Santa Cruz Avenue
Menlo Park, CA 94025

Pollution Contr. Res. Inst.
3875 Jackson, Suite 6
Riverside, CA 92503

Jan Pfuhl
3436 Pierce St. #3
San Francisco, CA 94123

Princeton Energy Group
729 Alexander Rd.
Princeton, N.J. 08540

Passive Solar Associates
P.O. 6023
Santa Fe, NM 87501

Rob Quigley, AIA
662 States St.
San Diego, CA 92101

Jeff Reiss & John Burton
4438 G Street
Sacramento, CA

Steven Robinson Associates
2114 Broadway
New York, N.Y. 10023

Mr. B.T. Rodgers
P.O. Box 2
Embudo, NM 87531

Mr. Kenneth Russell
San Joaquin Valley SEA
Cal State Univ. Fresno
Fresno, CA 93740

Jesse Savell
Passive Solar-Variant Homes
575 Birch Court, Suite A.
Colton, CA 92324

Professor Richard Schoen
Architecture & Urban Planning
University of California
Los Angeles, CA 90024

Santa Barbara Solar Energy Systems - Thomas Miller Assoc.
1702 San Andres St., #1
Santa Barbara, CA 93101

William A. Shurcliff
19 Appleton Street
Cambridge, Mass. 02138

Self-Help Enterprises
220 South Bridge St.
Visalia, CA 92377

Setter, Leach and Lindstrom
1011 Nicollet Mall
Minneapolis, Minn. 55403

Skytherm Processes & Eng.
2424 Wilshire Blvd.
Los Angeles, CA 90057

So. California Solar Energy Association
202 "C" Street
San Diego, CA 92101

The Solar Center
62 Townsend Street
San Francisco, CA 94107

Gary Starr
Natural Systems of Lake County
P.O. 185
Lower Lake, CA 95451

Jonathon Stoumen Architect
P.O. Box 87
Miranda, CA 95553

Sun House Design
Box 130
Occidental, CA 95465

Sunlight & Power Co.
115 Park Place
Point Richmond, CA 94801

Jay Sway
Geobuilding Systems
P.O. 1556
Hereford, TX 79045

Ray Sterling
The Underground Space Center,
Dept. of Civil & Mech Engineering
Univ. of Minnesota
Minneapolis, Minn. 55455

Sun Energy Builders
5838 Robertson Ave.
Carmichael, CA 95608

Norman Saunders
15 Ellis
Weston, Mass 02193

Tod Sloan (Savell Licensee)
Sloan & Assoc. Inc.
7248 Fair Oaks Blvd.
Carmichael, CA 95608

Tom Smith
Box 2356
Olympic Valley, CA 95730

Total Environmental
Action (TEA)
Harrisville, NH 03450

Karen Terry
Sun Up, Inc.
636 Camino Lejo
Santa Fe, NM 87501

Tackett-Way-Lodholz
Architecture, Planning, & Energy
3123 Buffalo Speedway Suite 400
Houston, TX 77098

Rob Thayer - Solar Access Planning
2321 Portage Bay
Davis, CA 95616

Malcolm Wells
P.O. Box 1149
Brester, MA 02631

D. Wright, AIA
Box 49
The Sea Ranch, CA

Steve Wolff
P.O. 1135
Berkeley, CA 94701

Charlie Wing
54 Cumberland St.
Brunswick, ME 04011

What-works
P.O. 588
Calistoga, CA 94515

Bill Wilson
2306 Reile
Escalon, CA 95320

Jim Zanetto
2310 Portage Bay #4
Davis, CA 95616

Zomeworks
P.O. 712
Albuqurque, NM 87103

DAYLIGHTING

Mr. J.W. Griffith
Consulting Engineer
3921 Caruth Boulevard
Dallas, TX 75225

Mr. W.C. Burkhardt
Rohm & Haas Company
Plastics Engineering Laboratory
Box 219
Bristol, PA 19007

Mr. Donald J. Vild
Libbey-Owens-Ford Company
1701 East Broadway
Toledo, OH 43606

Mr. Francis Clark
Lighting Services
P.O. Box 3009
Waterbury, CT 06705

Mr. David L. DiLaura
Smith, Hinchman & Grylls Assocs., Inc.
455 West Fort Street
Detroit, MI 49226

Mr. Ben H. Evans
College of Architecture
Virginia Polytechnic Institute
Blacksburg, VA 24060

Mr. Walkley B. Ewing
Sunscreen Company
25 Close Road
Greenwich, CT 06830

Mr. Michael G. Bitterice, Engineer
Technical Services, Flat Glass Div.
PPG Industrices, Inc.
One Gateway Center
Pittsburgh, PA 15222

Mr. Robert Ferrell
College of Environmental Design
University of California
Berkeley, CA 94720

Mr. H.F. Kingsbury
Pennsylvania State University
230 Hammond Building
University Park, PA 16802

Mr. J.B. Murdock
Department of Electrical Engineering
Kingsbury Hall
Unversity of New Hampshire
Durham, NH 03104

Mr. Steven Squillace
Smith, Hinchman and Grylls Assocs., Inc.
453 West Fort Street
Detroit, MI 48226

Jens Pohl
San Luis Obispo Solar Group
2065 McCollum
San Luis Obispo, CA 93401

Mr. Fred S. Dubin, President
Dubin-Mindell-Bloome Associates
42 West 39th Street
New York, NY 10018

Richard Stein and Partners, Architects
588 Fifth Avenue
N.Y., N.Y. 10036

Mr. Gershon Meckler
Gershon Meckler Associates
1150 - 17th Street, N.W.
Washington, DC 20036

Dr. John E. Flynn
Professor of Architectural Engineering
Pennsylvania State University
University Park, PN 16802

Mr. Richard E. Harmon
PPG Industries, Inc.
300 N. Lee Street, Suite 302
Alexandria, VA 22314

The Passive Solar Institute

The Passive Solar Institute was founded on November 1, 1978 to promote more widespread use of passive solar heating and cooling. Its primary goals are:

1. Information Delivery
The biggest obstacle to widespread use of passive solar heating and cooling is ignorance. The Passive Solar Institute will help deliver existing information and news of future developments to potential users rapidly and concisely. The primary mechanism for this transfer will be an annual Passive Solar Catalog with descriptions of techniques and access to materials, equipment, and consultants for Passive Solar Design. The Passive Solar Institute will also conduct workshops and classes on passive solar design.

2. Information Development
The existing "state of the art" is still rather primitive and much more basic research needs to be done. Limited private and government spending on research is retarding the rapid penetration of passive solar systems into the mass market. The Passive Solar Institute will identify areas where further research is needed and carry out basic research. In addition to in-house research the Passive Solar Institute hopes to sponsor basic research by other firms.

3. Product Development
The main thrust of the Passive Solar Institute will be in the identification and development of passive solar components. A series of Market Reports will identify needed components, their use, and design requirements for potential manufacturers. In addition, existing products that are or can be used for passive systems will be identified. The Passive Solar Institute will work closely with the manufacturers in developing marketing strategies for these components.

4. Planning for Energy Efficiency
The Passive Solar Institute will also be involved in land use planning for energy efficiency. Primary focus will be on planning for Solar Access but more broad concerns of energy conservation and efficient resource use will also be addressed.

5. Energy Policy
The Passive Solar Institute will also provide information on passive solar use for energy policy development at the local, state, federal, and international level. The tremendous potential for passive solar heating and cooling, passive hot water heating, day-lighting, and solar access planning needs to be carefully evaluated as part of any responsible energy plan. It is a case of almost limitless opportunity that is currently being ignored. Emphasis will be placed on mechanisms to remove the unwarranted subsidies for non-renewable fuels and reduce government involvement in the energy sector.

6. Profit
Although the concept of profit-making institutions has fallen out of favor in recent year, The Passive Solar Institute intends to make a consistent profit. Free enterprise and a profit provide the best incentive for doing a job well, on time, and under budget.

7. Demonstration
And finally, The Passive Solar Institute will help demonstrate the value of passive solar systems in both new and retrofit buildings. The Passive Solar Institute operates from a building with day-lighting, passive heating and cooling, and passive solar hot water. Future projects of The Passive Solar Institute will include retrofits, new residential and commercial building design, and hopefully side-by-side testing of different passive systems.

David A. Bainbridge
President
The Passive Solar Institute
2446 Bucklebury
Davis, Ca. 95616

TELL THEM YOU HEARD ABOUT THEM IN THE PASSIVE SOLAR CATALOG

Chapter 12. Passive Solar Use and Energy Independence

1. Introduction

As you have read through the Passive Solar Catalog you have seen many examples of how passive solar design works and why it makes sense. In most cases the use of passive solar design can reduce building energy use for space conditioning and water heating 50% or more. This Chapter looks at how Passive Solar Design can help make America more energy independent and why that is increasingly important.

2. Why bother?

While those in business and finance may already know why reducing America's use of non-renewable energy is important, many members of our society don't realize that we continue to import almost ½ of the oil we use. Here are some very strong reasons for reducing this profligate consumption of non-renewable fuels.

A. Freedom from Oil Blackmail

Our energy use has exceeded what we used before the great oil crisis of 1973. In fact, our energy situation has gotten worse. Oil imports make up an even larger share of our energy supply - and make us ever more susceptible to political pressure. As several countries demonstrated in 1973 such dependence may cause almost complete political amorality in return for oil. This is as dangerous as heroin addiction - both for the addict and the supplier.

B. Balance of Payments

There is a certain irony in this phrase, as in recent years, America's imports have consistently exceeded her exports. In September 1978 the trade imbalance was $1.69 billion and the total in 1978 may exceed $20 billion. This imbalance could be eliminated if we reduce the $45 billion we spend on oil imports every year.

This trade imbalance hurts the U.S. in many ways, including rapid increases in the prices of raw materials which are increasingly bought with recycled American oil dollars. For example, wood one of the fastest price gainers, is being heavily sold to Japan and paid for with oil dollars gained by selling industrial equipment to the oil producers. This hurts not only the home buyer and builder - but also the mill worker who is laid off because housing starts are down and raw lumber prices are up. This is a major factor in the 20% year increase in construction material prices.

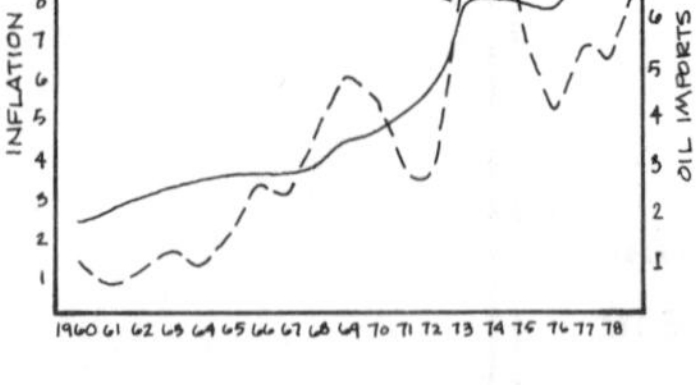

C. Inflation

The most insidious effect of relying excessively on foreign oil and capital intensive non-renewable energy sources is inflation. It may be going too far to say that inflation is entirely a result of bad energy policy - but it is at least partially responsible. Our reliance on foreign oil and the resulting imbalance of payments cheapen the dollar and make it less valuable both abroad and at home.

The government's reliance on and vast investments in non-renewable fuels have also helped deepen the Federal Budget deficit. Even a small portion of the $1 trillion dollars spent since 1945 in non-renewable fuels development programs and the $100 billion dollars per year spent directly subsidizing non-renewable fuels would have been sufficient to offset the total Federal deficit of $338 billion. The deficit is now increasing at the rate of almost $50 billion/year.

D. Better Investment

Passive solar design also makes much better economic sense than continued reliance upon non-renewable fuels. For example, assume we are building a house in the Central Valley of California. If, as is now current practice, it has 55 sq. ft. of glass (one patio door and two windows) facing west it will require one ton of air conditioning to keep the house cool. Because the air conditioner is only 70% efficient (EER=7), and production and distribution losses are incurred for electricity generation, it will require almost 2 kw of generating capacity at the peak period of electrical use to operate the air conditioner. The cost of the air conditioner and generating plant capacity total around $3,000. Alternatively, we could have spent $0 for reorienting the house, $20 for a rolldown bamboo shade, or $300 for shade screen. Which is the better investment?

E. Better Environmental Impact

Passive solar systems are also better because they have largely beneficial environmental impacts. They are in fact one of the only clean methods of "generating" energy. Every retrofit or new passive house frees as much as 3 kw of generating capacity for other uses. The use and production of the non-renewable fuels and some of the renewable fuels have serious adverse environmental impacts. Passive solar doesn't and as recent events such as the chemical contamination of Seveso, Italy have demonstrated, we would do well to minimize the potential for any serious environmental contamination.

Passive solar investment also makes jobs - a very positive social impact. As we said earlier, passive solar may create 10 times as many jobs as non-renewable fuels for a given investment. The jobs created from a large investment in Passive Solar could be used to offset the loss of jobs caused by increasing government cutbacks from tax initiatives like Prop. 13 in California.

F. More Efficient

Passive solar systems are also a better investment because they are more efficient. As Amory Lovin's so clearly stated it, "using electricity to heat water is like using a chainsaw to cut butter." Solar energy falls in usable quantities on almost every house in the U.S.

Passive solar systems are ideally suited to the demand for relatively low temperature (less than 200°F) heating demand for space heating, water heating, and some process heat. They can also eliminate the need for air conditioning. In a recent study of end use of energy in the U.S. almost 30% of the total energy use was found to be for demands which passive solar systems can meet.

G. Fast

It may take a utility 10-15 years to get a power plant designed, approved, and built. A passive solar system can be designed and built in three months (for a house), a week (for a hot water heater), or a few hours (thermal drapes or shutters). This is particularly important: when we decide we must change - we can do it in a matter of years, rather than decades.

H. Risk

Perhaps the most important reason for favoring passive solar design is the low element of risk involved. Passive systems have low risk from: supply disruption, mechanical failure, health hazards, and breakdown caused by natural disaster or terrorist activity. These risks are worth a closer look.

As long as the sun continues to shine passive solar systems will work. This is a considerably better bet than continued supplies of oil from the mid-east or even coal distribution within the U.S. As recent events have demonstrated, the rail systems may or **may not** be able to move coal to the power plants when it is needed. As the rail system continues to deteriorate this problem will become increasingly serious. A strike within the transportation or energy industry could easily disrupt the centralized non-renewable energy systems but it can't stop your passive solar heating system or hot water heater from working.

The smallness and simplicity of passive solar systems remove the element of risk from mechanical failure. If worse comes to worst and your hot water system fails (tanks will corrode, although less quickly than they do in a conventional heater) you can always go next door for a shower. But if a 2000 megawatt reactor goes down at the wrong time it may take the whole system with it, causing regional blackouts - and complete failure of conventional electrical heating systems.

The most favored option by many utilities has been nuclear generating plants. Although their safety record to date has been very good they are particularly sensitive to risk of health hazard because they involve such dangerous materials. When even a modest mechanical failure occurs it can cause the regulating agencies to shut all similar systems down until a specific mechanical defect is fixed. The limited number of types of systems has made this particularly troublesome and several cases like this have already occurred. In the event of a serious failure or accident public outcry might shut all of the reactors down - leaving everyone desperately short of electricity.

And finally, decentralized energy systems are much less likely to be damaged by natural disasters or terrorists. A large earthquake, such as the New Madrid, Missouri quake of 1812 would quite possibly destroy some of the generating capacity and much of the transmission and distribution system in a broad area. This might take weeks, months, or even years to repair. Yet passive solar systems will still be working in most of the houses in the affected area. House will stay warm (or cool) and as soon as water lines are repaired people can have hot water. The threat of terrorist acts is also increasing and centralized utility systems are very attractive targets. Passive solar systems are not.

Passive solar systems can provide the security of a comfortable home in an increasingly troubled world.

3. How Do We Get There From Here?

The simplicity and low cost of passive solar system will enable very rapid market penetration to occur if we end subsidies to non-renewable fuels, increase private and public investment in passive solar, and work hard

A. End Subsidies

As we suggested earlier we probably would still be using passive solar systems and other renewable sources of energy if the government subsidies for non-renewable fuels were not so high. Therefore, the first (probably futile) effort should be to reduce or eliminate the many subsidies for non-renewable fuels. These include: rolled in pricing, fuel tax credits, special depreciation allowances, the depletion allowance, credit for foreign taxes paid, etc. etc.

B. Increased Investment

We will also have to increase our investment in Passive Solar Systems. This will mean increased activity by: 1) the individual or family, 2) the builder, 3) busines and industry, 4) the utilities, and 5) governments. Th potential for passive development in these variou sectors is briefly described in this section.

1. The Individual or Family

Personal investment in passive solar design is already the major source of funding for development and installation of passive systems. People are wiser than the economists believe and are making the investment because it will save money and provide increased security. Passive solar systems are one of the best investments possible in an inflationary economy. As fuel costs soar, the low fixed cost of solar-heated homes will grow increasingly attractive.

The rapidity of change that can be expected was demonstrated in Vermont after the "oil crisis." Between 1974-1976 forty percent of the houses in Vermont were back fit with wood stoves with no state or Federal intervention. Another example of the changes that can occur is the remarkable increase in home gardening between 1971-1976. The percentage of households growing some or all of their own food rose from 20-50%. We can do it, if we want to.

2. The Builders

The houses of American are predominantly designed and built by small builders. It is an area of American business where independence and hard work are still the rule - rather than the exception. And it is a business that is beginning to move toward passive design. The California Building Industry Association was one of the first groups to really understand the implications of the continuing reliance on foreign oil and non-renewable fuels. In a resolution passed in August, 1978 they concluded: "the California Building Industry Association encourages home builders in California to incorporate energy conservation and passive solar design in houses . . .".

They are currently sponsoring a Building Seminar series which features both Passive and Active solar design. One of the faculty members for the series, Bob Kuntz of CBIA, had this to say, "*if we build well and reduce the energy demand we can reduce the need for new power plants. We are building houses which people will be living in 40-50 years from now so they had better be good.*"

The building industry can make almost all of the new housing built by 1980 passive solar.

3. Business and Industry

Business and industry can also move fairly quickly to more widespread use of passive solar design. Pacific Telephone's recent commitment to hold energy use constant despite increases in system size and capacity is a noteworthy example. Commercial buildings are particularly attractive for passive systems because even in cold areas internal heat gain and fairly limited solar glazing can meet the full heating demand. The potential savings from good solar control programs, with shadescreen and roof coatings, are also staggering and should lead to fast and highly profitable growth in those industries.

The Passive Solar Institute will devote much of its effort to stimulating activity in Passive Solar market development. The demand is there but in many cases the installers or the products aren't. Business and industry have fallen down in this regard and their failure does not bode well for our future. Research and development have suffered from increasing corporate complacency. Companies should be investing more both in their own research and also in independent research. It has become common to "let Uncle Sam do it," even though the Federal government has often proved unwise in choosing research projects.

4. Utilities

The utilities can also play an important role in passive solar development. In particular they must do better advance planning than they have to date. Risky and economically questionable programs have been adopted too frequently without consideration for how the government, the public, or the environment will be affected. Continued investment in very large nuclear or coal plants, despite their lower reliability than smaller plants is one example of a breakdown in their planning.

In part, this misguided policy is a result of the utilities unique position somewhere between being an industry or a governmental body. Regulated and protected they suffer from the worst effects of both. Profits are often guaranteed by regulating agencies, based on investment. Hence, it is reasonable to invest in big plants because they cost more. And while they can't lose - their customers certainly can! They are also in many cases overregulated and deserve full credit for continuing to produce power in what is often a nightmare of endless review and red tape.

If the utilities weren't so oddly constituted they would have long ago invested in energy conservation, cogeneration, passive solar design, and renewable fuels. A small utility in Vermont has demonstrated the viability of renewable fuels - without Federal support or public subsidy. By retrofitting an older coal fired plant to burn wood Burlington Electric has reduced their sensitivity to supply disruption, reduced cost of energy generation, and proved the concept of using cull wood to generate electricity. By 1982 Burlington Electric will meet half of its energy demand using wood.

"We'll be spending $30,000 a day for fuel and all that money will be staying right here in Vermont. Strengthening our local economy."

Robert Young
Burlington Electric General Manager

The utilities can also play an important role in financing renewable energy and energy conservation installations on homes. If the rules and incentives under which they operate are modified only slightly they could and would provide the money needed for retrofits, recovering their investment in the monthly utility bills.

As the many legitimate environmental, social, and economic concerns make adding increased capacity ever more difficult the utilities will have to take a more aggressive solar/conservation stance or face rolling brownouts and blackouts. They may find that it can be a time of growth and challenge as interesting as the building boom after World War II. The Passive Solar Institute looks forward to working with the utilities during this transition.

5. Governments

The Federal, State, and local governments could also do much to implement passive solar design. Rather than mandating solar waterheating, like San Diego County recently did, they should remove the subsidies for renewable fuels, remove regulatory barriers, provide good information, let the market forces work, and set a good example by retrofitting their own buildings.

While the subsidies for non-renewable fuels are being dismantled governments should also invest in passive solar design. It may be the best defense investment we as a nation can make. Only $10 billion a year could retrofit almost a million houses per year and support a very strong research program.

And rather than overregulation and swamping the market with rule changes governments should set long term standards (if they set any at all) and stick to them. Better yet, they should set product specifications instead, and then buy a few hundred million dollars worth of passive solar products a year for retrofitting. Those companies that meet the specifications will prosper and build ever better products.

3. What We Should Aim For - Negative Energy Growth

A comprehensive energy program for the U.S. is far beyond the scope of this chapter in the Passive Solar Catalog. Yet it is not difficult to suggest what the goals should be and what we could achieve if we tried. Simply stated the goal should be the transition over the next twenty years from non-renewable fuels to renewable fuels. An immediate goal should be a negative growth rate in energy use (we might settle for zero growth) and a reduction in oil imports from 50% of total use to less than 25%.

A look at the impact of passive solar design on total energy use will demonstrate that these goals are realistic. Although, passive solar design will be one of the easiest transitions it is not that much harder to make equal or greater savings in other areas.

The current housing stock in the U.S. includes 80 million units. Another 2 million are built every year. At 2.5% growth per year this is slightly ahead of the population growth rate - but not much. We can, therefore maintain zero energy growth if we make sure that all the new houses built feature passive solar orientation and retrofit 2 million houses per year, assuming a 50% reduction in energy use for passive solar homes.

To do this would require an investment of only about $3 billion a year. This is about 1/25 the amount being spent each year on new power plants! If we make a more credible effort to retrofit houses we can do much better. For example, if we spend $12 billion/year we might retrofit nearly 10 million houses/year. This could save an additional 750 billion BTU/year. By 1985 we could retrofit almost all of the houses in the U.S. and be saving on the order of 3 quadrillion/BTU a year. Our survival as a free and independent nation may depend on our doing no less.

Similar savings can be realized in other sectors of society. The major uses are industry, space conditioning, transportation, and energy generation and transportation. Almost 50% of total use is involved in building construction, maintenance, operation, and land use planning and we have shown the type of savings that can be made in that sector. Industry and transportation could reduce their use by 50% with only slightly more difficulty.

The City of Davis, California has demonstrated the feasibility of achieving negative energy growth. With a relatively modest program, begun in 1972, they have reduced total use of natural gas from a high in 1975 of 1,904,953 mcf to 1,477,578 in 1977. During the same period total electrical use is down from a high in 1976 of 132 million kwh to 129 million kwh in 1977. A significant reduction in use per customer has been responsible for this negative rate of growth in a period of considerable physical growth in the city.

Per customer use of natural gas has dropped from a high of 190 mcf in 1975 to only 135 mcf in 1977. Residential use peaked in 1971 at 115 mcf and has now dropped to 82 mcf, a reduction of nearly 30%. Per customer use of electricity peaked in 1972 at 11,940 kwh and has now dropped to 9,853 kwh. Residential use has dropped from a high of 7,300 kwh in 1973 to 6,320 kwh in 1977, a 14% reduction.

Davis has managed to accomplish this without coercion, overregulation, or major investment. The cost of the city program to date is probably less than $10 per capita. Emboldened by their success they have now set a goal of reducing total energy use 50% by 1986. We as a nation can do no less.

Further Reading

Why Bother?

Amory Lovins (1977) **Soft Energy Paths**, Ballinger.

Wilson Clark (1975) **Energy for Survival**, Anchor Press.

Denis Hayes (1976) **Energy: The Case for Conservation**, Worldwatch Institute.

Emilio Varanini, Robert Boyd, et. al. (1977) **Toward A California Energy Policy**, Vol. I of the Biennial Report, 1111 Howe Ave., Sacramento, CA 95825 $2.50.

Herman Daly (1973) **Towards A Steady State Economy**, W.H. Freeman.

Robert Heilbronner (1974) **An Inquiry into the Human Prospect**, Norton, N.Y.

Meadows, D. et. al. (1972) **The Limits to Growth**, Potomac Associates, Washington.

Komanoff, C. (1976) **Power Plant Performance**, Council on Economic Priorities, N.Y.

Oregon Office of Energy Research and Planning (1975) **Transition**, Salem, Ore.

Ivan Illich (1974) **Energy and Equity**, Perennial.

Barbara Ward and René Dubos (1972) **Only One Earth**, Penguin.

Paul and Ann Ehrlich (1972) **Population, Resources, Environment**, W.H. Freeman.

John Fuller (1978) **The Poison that Fell from the Sky**, Random House, N.Y.

Paul Erdman (1976) **The Crash of '79**, Simon and Schuster.

Bruce Hannon (1976) "Energy and Labor Demand", Center for Advanced Computation, Urbana, Illinois.

Ayn Rand (1957) **Atlas Shrugged**, Random House.

Wendell Berry (1977) **The Unsettling of America**, SCBooks.

M. Willrich, ed. (1973) **International Safeguards and Nuclear Industry**, John Hopkins Press.

P. Morrison and P. Walker (1978) "A New Strategy for Military Spending" **Scientific American**, Oct.

How Do We Get There From Here?

Marshall Hunt and David Bainbridge (1978) "The Davis Experience" **Solar Age**, May

Amory Lovins (1977) **Soft Energy Paths**, Ballinger.

Peter Van Dresser (1972) **A Landscape for Humans**, Lighting Tree Press, Santa Fe.

Richard Stein (1978) **Architecture and Energy**, Anchor Press/Double Day.

Victor Papanek (1972) **Design for the Real World**, Bantam Books.

Helen and Scott Nearing (1970) **Living the Good Life**, Schocken.

Ernest Callenbach (1975) **Ectopia**, Banyan Tree Books.

Fritz Schumacher (1973) **Small is Beautiful**, Harper and Row, N.Y.

W.H. Harman (1976) **An Incomplete Guide to the Future**, SAA, Palo Alto

Leckie, J. Masters, G., Whithouse, H. and Young, L. (1975) **Other Homes and Garbage**, Sierra Club Books.

Chris Williams (1974) **Craftsmen of Necessity**, Vantage.

Hammond, J., Kopper, B., Bainbridge, D. et. al. (1977) **The Davis Energy Conservation Report**, Living Systems, Winters, CA.

Bainbridge, D. and Hammond, J. (1976) **Planning for Energy Conservation**, Living Systems.

Richard Congdon, Ed. (1977) **An Introduction to Appropriate Technology**, Rodale Press.

Solar Age (1978) "Solar and Utilities" the December issue.

David A. Bainbridge

NOTES:

About the Author:

David Bainbridge was born in 1948 in Schenectady, N.Y. At the time his parents lived in a log cabin nestled in the woods by a beautiful mountain stream and the mountains and wilderness have continued to play an important role in his life. Family moves and travel have introduced him to the North Cascades, Canadian and Colorado Rockies, and the Sierra Nevada.

After graduating from U.C. San Diego with a major in earth sciences and a minor in surfing he worked for the San Diego Environmental Development Agency in a successful effort to save the coastal lagoons from imminent development. This project led to a Masters degree from U.C. Davis in Ecology with emphasis on ecology, planning, and environmental geology.

While going to school in Davis he worked with the state Joint Committee on Environmental Quality and formed an environmental consulting firm which specialized in Bikeway Planning, Environmental Impact Analysis, and Environmental Planning and Management.

He retired after two hectic and rewarding years to help his parents build a homestead in Cortez, Colorado. This involved completely dismantling and recycling a dilapidated house, including the nails, into an energy conserving house.

On his return to California he worked for the Association of Bay Area Governments on an innovative Land Capability study. At the same time he began the restoration of an 1892 Victorian house in a small town near Sacramento that was to take two years. After completing the ABAG report he began what was to become two years of very interesting and productive work with a solar design group in Winters, California.

These two years with Living Systems included research, writing, photography, and design of passive solar and energy conservation projects. David was in charge of the Planning for Energy Conservation studies in Davis, Indio, and Sacramento County. He also was involved in natural lighting studies for the State Office Building, the Energy Conservation Building Code Workbooks for Indio and Sacramento, and the first drafts of a report on the challenge of planning for Solar Access.

He retired to devote more time to completing his passive solar home in Davis and to his writing. Unfortunately, retirement lasted only a month before an urgent call put him to work on the development of the Passive Section of California's Solar Tax credit at the Solar Office of the California Energy Commission. Other reports and studies followed at regular intervals.

He resigned from the Energy Commission to form the Passive Solar Institute and promote the widespread and rapid transition to passive solar utilization. He has many reports underway in addition to the Passive Solar Catalog.

After finishing his house in Village Homes in January 1978 he became more involved in the affairs of this remarkable 70 acre solar subdivision. He has served on the Board of Directors, the Small Farms review board, and has written a book about the development which will be published by Rodale Press early in 1979.

In addition to his work on the Passive Solar Institute he does freelance writing, photography, energy consulting, lecturing, and concrete and carpentry work for friends building in Village Homes. On his time off, which is all too rare, he enjoys backpacking, skiing, surfing, canoeing, playing guitar, and writing poetry.

photo: Bob Sculley

Over 300 products for Passive Design

More than 150 Passive Solar Consultants

How to: Build passive Solar houses, water heaters, insulated shutters, and much much more.

Marshall Hunt, California Energy Commission:

A must for anyone considering building or buying a new home!

Virginia Thigpen, Contractor:

This book should be on every builders desk (... board). The time and energy it will save will pay for ... hundreds of times over.

To order more copies of The Passive Solar Catalog:

Send check or mone... ...ail oversea...

...avis, CA 95616

The First Passive Solar Catalog

ISBN 0-9334...